T0140203

Monographs in Theoretical Computer Science
An EATCS Series

More information about this series at http://www.springer.com/series/776

Sergey Kitaev • Vadim Lozin

Words and Graphs

Sergey Kitaev
Department of Computer
and Information Sciences
University of Strathclyde
Glasgow, UK

Vadim Lozin
Mathematics Institute
University of Warwick
Coventry, UK

Series Editors

ISSN 1431-2654 ISSN 2193-2069 (electronic)
Monographs in Theoretical Computer Science. An EATCS Series
ISBN 978-3-319-35669-3 ISBN 978-3-319-25859-1 (eBook)
DOI 10.1007/978-3-319-25859-1

Springer Cham Heidelberg New York Dordrecht London

Softcover re-print of the Hardcover 1st edition 2015

Printed on acid-free paper

Springer International Publishing AG Switzerland is part of Springer Science+Business Media (www.springer.com)

Foreword

This excellent book by Sergey Kitaev and Vadim Lozin is the first book devoted to the important topic of word-representable graphs. It is a comprehensive presentation of the state of the art, gathering, unifying and summarizing the large body of results that have been developed over recent years. It raises new open problems to be further explored by researchers, and then introduces additional connections between words and graphs, which are worth studying further.

Let w be a word over an alphabet V and suppose that x and y are two distinct letters in w. We say that x and y alternate in w if after deleting in w all letters but the copies of x and y we either obtain a word $xyxy\ldots$ (of even or odd length) or a word $yxyx\ldots$ (of even or odd length). If x and y do not alternate in w, we say that these letters are non-alternating in w. A graph $G = (V, E)$ is word-representable if there exists a word w over the alphabet V such that letters x and y alternate in w if and only if $(x, y) \in E$ for each $x \neq y$.

The family of word-representable graphs generalizes several well known classes including comparability graphs, 3-colorable graphs, circle graphs and graphs of vertex degree at most 3. But not all graphs are word-representable, making them a family of interest. Indeed, the authors suggest a variety of problems and directions for further research, some open for several years and others published in this book for the first time.

The book offers much more than an introduction to the theory of word-representable graphs. It presents a collection of various interrelations between words and graphs in the literature. These include discussion of well known notions such as Prüfer sequences, Gray codes or de Bruijn graphs, and some less well known relations between graphs and words, such as permutation graphs, polygon-circle graphs, word-digraphs and path-schemes.

In conclusion, "Words and Graphs" is the most valuable and definitive source volume for entering this interesting area of research. It is enjoyable to read, and will become very useful for both learning and reference. Its challenging problems give insight into the progression of research over the last decade, and will keep a new generation well occupied for the coming decade.

Haifa, June 2015

Martin Charles Golumbic

Preface

In 1918, Heinz Prüfer [120] discovered a fascinating relationship between labelled trees with n vertices and sequences of length $n-2$ made of the elements of the set $\{1, 2, \ldots, n\}$. This relationship is, in fact, a bijection, i.e. a one-to-one correspondence between trees and sequences, and it allowed Prüfer to prove Cayley's formula about the number of n-vertex labelled trees. The Prüfer sequence is a classical example showing the importance of words for graph enumeration. More importantly, with the advent of the computer era representing graphs by words became crucial for storing graphs in computer memory. Words have also been used to reveal and describe various useful properties of graphs, such as classes where many difficult algorithmic problems become easy, or classes that are well-quasi-ordered by the induced subgraph relation. On the other hand, graphs have frequently been exploited to study various properties of words and related combinatorial structures, such as permutations.

Since the discovery of the Prüfer sequence, the interplay between words and graphs has repeatedly been investigated in both directions. One of the most recent findings in this area is the notion of *word-representable graphs*. This is a common generalization of several well-studied classes of graphs, such as circle graphs, comparability graphs, 3-colourable graphs and graphs of degree at most 3 (also known as subcubic graphs). The invention of word-representable graphs became the inspiration for writing this book. On the other hand, it motivated us to look at various other important relationships between words and graphs. In the first part of the book, we report, in a comprehensive way, the state of the art on word-representable graphs and give a brief tour over related graph classes. In the second part, we explore many other connections between words and graphs. In no way is our description of these connections comprehensive or complete. Rather, it is an invitation to an area that faces many great challenges and offers the prospect of many great discoveries.

The book is organized as follows.

- In Chapter 1 we provide a quick introduction to the world of word-representable graphs, which is the main object of our interest in this book, and give a few motivating points for our study of these graphs.

- In Chapter 2 we discuss hereditary classes of graphs. The reason for our rather thorough discussion is that the class of word-representable graphs is hereditary and it generalizes several important representatives of this family. Thus, knowing general results and approaches to tackle problems for hereditary classes of graphs can be useful in dealing with word-representable graphs. In particular, the only enumerative result for word-representable graphs known to date was obtained by means of asymptotic enumeration developed for hereditary classes.

- In Chapters 3–5 we provide a comprehensive introduction to, and overview of known results in, the theory of word-representable graphs. A proof of each result is presented along with a reference to the original source.

- In Chapter 6 we discuss a generalization of the notion of word-representable graphs, namely that of u-representable graphs, where u is a binary word over $\{1, 2\}$. Our word-representable graphs are 11-representable in the new terminology. The focus of the chapter is the study of 12-representable graphs, where many interesting properties of these graphs are established; all the proofs are provided in a self-contained manner. We also show that *any* graph is u-representable assuming that u is of length at least 3.

- In Chapter 7 we suggest a variety of problems and directions for further research in the theory of word-representable graphs. These range from problems open for several years to those published in this book for the first time. The chapter also contains a section dedicated to possible approaches to tackle problems on word-representable graphs.

- In Chapters 8 and 9 we discuss interrelations of words and graphs in the literature by means other than word-representability. This includes a variety of topics without a common thread. We discuss not only well known notions, such as Prüfer sequences, Gray codes or de Bruijn graphs, but also less well known objects, such as permutation graphs and polygon-circle graphs, and structures essentially unknown to the general audience, such as word-digraphs and path-schemes.

- Appendix A contains, normally standard, definitions in graph theory that are used in the book, while in Appendix B we provide a few basic definitions in algebra, analysis and combinatorics that are used in the book. Note that many other (graph-theoretic) definitions are incorporated in other chapters, usually at the places where we need them.

Chapters 2–5 contain exercises and solutions to selected problems, making this book suitable for teaching purposes. The book will be of interest to researchers, graduate students and advanced undergraduate students with interests in graph theory, combinatorics (on words) and discrete mathematics.

Acknowledgments

The main focus of this book is the theory of word-representable graphs, and we would like to thank all the researchers contributing to the development of the theory. Special thanks go to Magnús M. Halldórsson and Artem Pyatkin for coming up with key results in the area, which influenced directions of further research in the field. It should also be noted that the theory would not be possible had Steven Seif not approached the first author of the book with a combinatorial problem in algebra, which resulted in the appearance of the prototype of the notion of a word-representable graph.

Also, we are thankful to the people who worked on creating software to work with word-representable graphs. These are Özgür Akgün, Herman Z.Q. Chen, Ian Gent, Marc Glen, Christopher Jefferson, Alexander Konovalov and Steve Linton.

Further, we would like to thank Manda Riehl for proofreading the first two chapters in the initial draft of the book, and providing many useful comments.

The first author would like to express his gratitude to Ian Ruthven for allocating extra time to work on this book. He is also grateful to his family for constant support, especially to his sons, Daniel and Nicholas Kitaev, for inspiration.

The second author is grateful to his wife, Irina Lozina, for encouragement, both by words and graphs, as graphs are nothing but relations.

Glasgow, September 2015 Sergey Kitaev

Coventry, September 2015 Vadim Lozin

Notation

$\lvert A \rvert$	number of elements in a set A
$[A]$	free commutative monoid on A; see Definition B.1.13
A^*	the set of all words over an alphabet A
$A - B$	removing from a set A the elements from a set B
$Alt(w)$	alternation word digraph for a word w; see Definition 3.6.3
$Alt_1(w)$	induced subgraph of $Alt(w)$ on singletons; see Definition 3.6.5
$A(w)$	set of letters occurring in a finite word w; see Definition 3.0.1
B_n	nth Bell number; see Definition B.2.4
$\mathbf{B}_2^1$	Perkins semigroup; see Definition 3.6.10
$c(G)$	supplement of a graph G; see Definition 6.1.13
$\mathcal{C}(k)$	class of graphs with clique-width at most k; see Theorem 9.5.3
C_n	cycle graph on n vertices; see Definition A.1.9
co-G	complement of a graph G; see Definition A.1.2
$c(w)$	complement of a word w; see Definition 6.1.10
$cwd(G)$	clique-width of a graph G; see Definition 9.5.1
$DC(P)$	uniform double caterpillar with spine P; see the proof of Theorem 6.4.4
D_k	set of graphs of degree at most k; see Subsection 2.2.5
$E(G)$	set of edges of a graph G
$\mathcal{E}_{i,j}$	graphs whose vertices can be partitioned into at most i independent sets and j cliques; see Section 2.3
$e(w)$	evaluation of w under e; see Definition 3.6.16
$(f_n^{\mathbf{M}})_{n\geq 1}$	free spectrum of a finite monoid $\mathbf{M}$; see Definition 3.6.21
$F(n)$	nth Fibonacci number; see Definition B.2.1
$Forb(X)$	set of all minimal forbidden induced subgraphs for a hereditary class X; see Section 2.1
$Free(M)$	graph class with no graph from M as an induced subgraph; see Section 2.1
$\overline{G}$	complement of a graph G; see Definition A.1.2
$\mathrm{Geom}(M)$	geometric grid class of a matrix M; see Subsection 8.3.2
$G(\mathcal{P}, w)$	letter graph of w; see Section 8.2
$G(w)$	graph represented by a word w; see Definition 3.0.9
G_w	word-digraph associated with a word w; see Definition 8.7.1

$G - v$	removing a vertex v from a graph G
$H_{k,k}$	crown graph; see Definition 4.2.4
$I(G)$	set of all independent sets of G; see Section 9.4
$\kappa(G)$	size of a maximal clique in G
K_n	complete graph on n vertices; see Section A.1
$K_{n,m}$	complete bipartite graph on $n+m$ vertices; see Section A.1
L_n	ladder graph; see Definition 5.2.14
$\mathcal{M}(\ell, n)$	word-digraph family; see Section 8.7
$\left\{ {n \atop k} \right\}$	Stirling number of the second kind; see Definition B.2.3
O_n	edgeless graph on n vertices; see Section A.1
$\mathcal{P}_n$	graph of overlapping permutations; see Definition 9.3.6
P_n	chordless path on n vertices; see Appendix A
$P(T)$	Prüfer sequence of a tree T; see Definition 8.1.1
$p(w)$	initial permutation of w; see Definition 3.2.11
$\text{red}(G)$	reduction of a graph G; see Definition 6.1.7
$\text{red}(w)$	reduced form of a word w; see Definition 6.1.3
Pr_n	prism on n vertices; see Definition 5.2.12
$\mathcal{R}(G)$	representation number of a graph G; see Definition 3.3.3
$\mathcal{R}_k$	class of graphs with representation number k; see Definition 3.3.3
$R(n, m)$	Ramsey number; see Definition A.1.12
$r(w)$	reverse of a word w; see Definition 3.0.12
S_n	star tree; see Definition A.1.11
$\mathcal{S}_n$	set of permutations of length n; see Definition B.1.2
$u \to v$	edge directed from u to v in a digraph
$u \approx_{\mathbf{S}} v$	$\mathbf{S}$-equivalence of u and v under a semigroup $\mathbf{S}$; see Definition 3.6.18
$U_X^{(n)}$	n-universal graph for X; see Subsection 8.3.3
$V(G)$	set of vertices of a graph G
$V_1(w)$	set of variables occurring once in a finite word w; see Definition 3.6.11
$Var(w)$	set of variables occurring in a finite word w; see Definition 3.6.11
w_B	removing the letters in $A(w) \backslash B$ from w; see Definition 6.1.1
W_n	wheel graph on $n+1$ vertices; see Definition 3.5.1
WQO	well-quasi-ordering; see Section 8.2

Contents

Chapter 1

Introduction

The core of our book is the theory of word-representable graphs, a young but very promising research area lying on the boundary of graph theory and combinatorics on words. This theory is discussed in Chapters 3–5, where we provide all necessary definitions and examples, as well as a proof of every statement we make along with references to the original sources.

In Chapter 6 we consider a generalization of the theory of word-representable graphs to that of pattern-matching representable graphs. Again, in Chapter 6 we aim to be comprehensive with respect to the objects we discuss. Further, Chapter 7 provides many research directions, sometimes from unexpected angles, related to word-representable graphs and their generalizations.

We are not so thorough in the rest of the book, comprising Chapters 2, 8 and 9. In particular, Chapter 2 offers a short tour of the world of hereditary classes of graphs to reflect the fact that the class of word-representable graphs is hereditary. Knowledge about hereditary classes of graphs in general has proved to be useful in asymptotic enumeration of word-representable graphs. It is conceivable that results over other hereditary classes of graphs could be useful in better understanding word-representable graphs.

We see the topic of hereditary classes of graphs as being especially important for our book, and we provide exercises for it in Chapter 2. We also provide exercises in Chapters 3–6 for our main objects of interest, thus making the book suitable for teaching purposes.

Chapters 8 and 9 contain a collection of interrelations between words and graphs in the literature. This collection is in no way comprehensive, but it contains many interesting items ranging from rather well known ones (like *Prüfer sequences*, *Gray codes* and *de Bruijn graphs*) to ones unknown to a general audience (like the *snake-in-the-box problem*, *path-schemes* and *word-digraphs*).

The rest of this chapter is devoted to our main object of interest, the class of word-representable graphs.

A graph $G = (V, E)$ is *word-representable* if there exists a word w over the alphabet V such that letters x and y alternate in w if and only if $xy \in E$, that is, x and y are connected by an edge, for each $x \neq y$. For example, the 4-cycle labelled by 1, 2, 3 and 4 in a clockwise direction can be represented by 13243142, and this word is called a *word-representant* for the cycle. Relations, to be established in this book, between word-representable graphs and some other classes of graphs are presented in Figure 1.1, which was essentially built by Magnús M. Halldórsson.

Word-representable graphs are also known in the literature as *representable graphs* or *alternation graphs*.

1.1 Basic Questions about Graphs Representable by Words

A number of basic questions one can ask about word-representable graphs are as follows.

- **Are all graphs word-representable?** In fact, no, although all graphs on at most five vertices are word-representable. Examples of non-word-representable graphs can be found in Sections 3.4 and 3.6.
- **How to characterize those graphs that are word-representable?** In Chapter 4 we discuss a characterization of word-representable graphs in terms of so-called *semi-transitive orientations*. This characterization allows us to solve the problem of deciding whether or not a given graph is word-representable by graph-theoretical means.
- **How many word-representable graphs are there?** Exact enumeration seems to be a very difficult problem, possibly comparable with such a problem as enumeration of partially ordered sets. However, one can employ general results on *hereditary classes of graphs* (the class of word-representable graphs is hereditary, which was first mentioned in [92]) to find the asymptotic growth in question; we discuss this in Section 5.3.
- **What is the minimum length of a word-representant for a given word-representable graph?** In Section 4.2 we will see that if a graph on n vertices is word-representable then one needs at most $2n$ copies of each letter

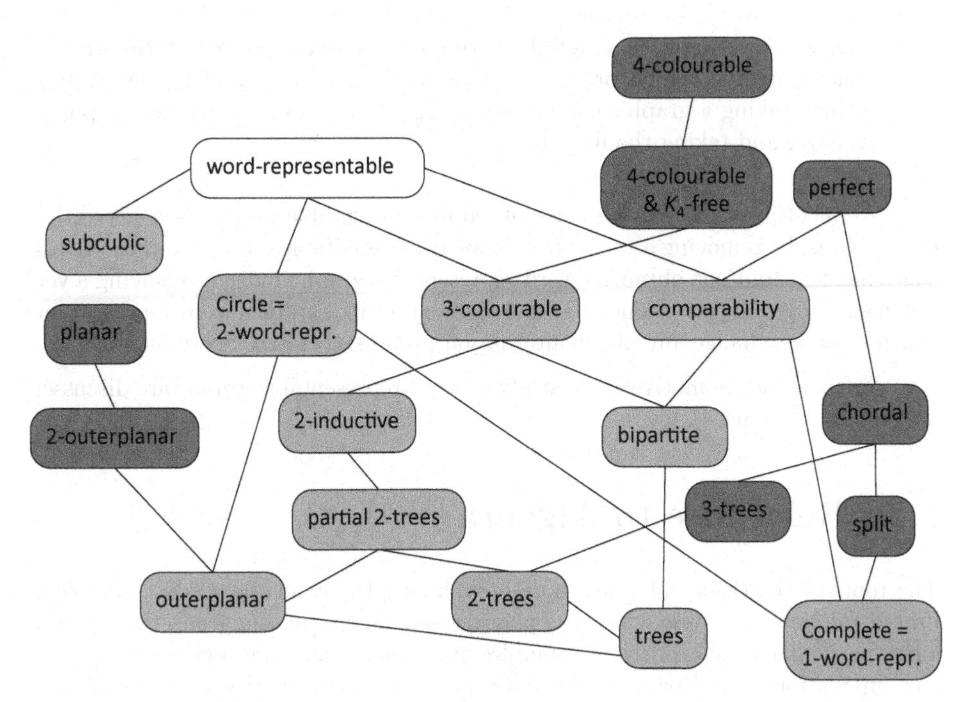

Figure 1.1: Relations between graph classes.

to represent the graph. There exist graphs requiring around $n/2$ copies of each letter to be represented, which we discuss in Subsection 4.2.1. However, we are not aware of any graphs requiring longer word-representants.

- **How hard is it to decide whether a graph is word-representable or not?** In Subsection 4.2.2 we will learn that it is an NP-complete problem to recognize whether a given graph is word-representable. A summary of complexity results known to us so far is presented in Subsection 4.2.3.

- **Is there a correlation between the chromatic number of a graph and its word-representability?** In Section 4.3 we will see that if the chromatic number of a graph is at most 3, then it is word-representable. On the other hand, for larger values of the chromatic number, the graph may or may not be word-representable.

- **Which graph operations preserve word-representability?** In Section 5.4 we shall see that some operations on graphs preserve the property of word-representability, while some other operations do not. Operations of the former type include gluing two graphs at a vertex, connecting two graphs by an edge, replacing a vertex with a module, taking the Cartesian product of two graphs, and taking the rooted product of two graphs; operations of the latter type include taking a graph's complement, edge contraction, gluing two graphs at an edge, and taking the line graph.

As mentioned above, the class of word-representable graphs is a hereditary class, that is, by removing a vertex from a word-representable graph together with all edges incident to it one obtains a word-representable graph. Indeed, removing a vertex in such a graph corresponds to removing a letter in a graph's word-representant, which does not change any of the alternation properties of other pairs of letters.

A few points to motivate the study of word-representable graphs are discussed in the next sections.

1.2 Relevance to Algebra

The roots of the theory of word-representable graphs are in the study of the celebrated *Perkins semigroup* which has played a central role in semigroup theory since 1960, particularly as a source of examples and counterexamples, and which we discuss in Section 3.6. However, the most interesting aspect of word-representable graphs from an algebraic point of view seems to be the notion of a semi-transitive orientation. In Chapter 4 we will see that a graph is word-representable if and only if it admits a semi-transitive orientation. Such orientations are a generalization of transitive orientations, and thus they are a generalization of partial orders, a very important structure in many contexts.

1.3 Importance for Graph Theory

Figure 1.1 gives a "proof without words" that word-representable graphs are indeed important from a graph-theoretical point of view, since they generalize several classes of graphs. The most striking facts are that word-representable graphs generalize

- *circle graphs*, which are 2-word-representable graphs in our language, while we also deal with k-word-representable graphs for any fixed $k \geq 1$ (see Sections 3.2 and 5.1);
- 3-*colourable graphs*. Indeed, all 3-colourable graphs are word-representable as shown in Section 4.3, while a word-representable graph on n vertices can have any chromatic number from 1 to n. The last statement follows from the

fact that any complete graph is word-representable (by a permutation). For another example, the union of a clique of size k and an independent set of size $n-k$ is word-representable (such a graph is a circle graph and it can easily be represented using two copies of each letter; see Section 5.1);

- *comparability graphs*, which in our language are precisely permutationally representable graphs (see Section 3.4).

1.4 Relevance to Computer Science

The study of word-representable graphs is interesting from an algorithmic point of view, leading to the following question. In which situations can knowing that a graph is word-representable, and possibly having a word that represents it, help to solve some problems on the graph that are hard to solve otherwise? For example, we will see in Theorem 3.4.14 that the Maximum Clique problem is polynomially solvable on word-representable graphs, while this problem is generally NP-complete. Having said that, we note that many classical optimization problems are NP-complete on word-representable graphs, as listed in Table 4.1.

1.5 Links to Combinatorics on Words

Word-representing a given graph is all about finding a word satisfying the right alternation properties for each pair of letters, and thus we immediately enter the domain of combinatorics on words, dealing with strings of symbols and their properties. We note that there are other instances in the literature where representing graphs by words has been considered. For example, *polygon-circle graphs*, being the intersection graphs of polygons inscribed in a circle, can be defined by words in which the existence of an edge is determined by the respective pair of letters alternating *at least once* as opposed to the strict alternation required in the case of word-representable graphs; we consider polygon-circle graphs briefly in Section 8.6. Another example is so-called *word-digraphs*, which are discussed in Section 8.7.

1.6 Relevance to Scheduling Problems

Inspired by the links between robotic scheduling and sequences satisfying certain alternation conditions described in [70] by Graham and Zang, Halldórsson, Kitaev and Pyatkin observed in [74] that word-representable graphs may find applications in scheduling problems. While this direction of research still requires much effort to become useful in practice, here are some ideas beginning with the example appearing in [74].

Consider a scenario with n recurring tasks and requirements on alternation of certain pairs of tasks, which captures typical situations in periodic scheduling, where there are recurring *precedence* requirements. For example, the following five tasks may be involved in the operation of a given machine: 1) Initialize controller, 2) Drain excess fluid, 3) Obtain permission from supervisor, 4) Ignite motor, 5) Check oil level. Tasks 1 & 2, 2 & 3, 3 & 4, 4 & 5, and 5 & 1 are expected to alternate between all repetitions of the events. One possible *task execution sequence* that obeys these recurrence constraints is 3 1 2 5 1 4 3 5 4 2.

Another scenario is building a conveyer by placing several copies of n types of robots in a line and respecting a given set of requirements, where, apart from the first cycle, robot i cannot do its job before robot j has completed its job, and vice versa, for some pairs (i, j), $1 \leq i < j \leq n$.

Alternation requirements on pairs of tasks can be represented by a graph whose vertices are labelled by tasks: there is an edge between two vertices if and only if the corresponding tasks alternate. On the other hand, execution sequences of recurring tasks can be viewed as words over the alphabet V, where V is a set of tasks. Not all requirements on pairs of tasks can be realized by an execution sequence, though any requirements on at most five or fewer tasks can be realized. If there exists an execution sequence realizing given alternation requirements, the corresponding graph is word-representable in our sense. The main question here is whether or not we can realize a given specification of task alternations.

Another example of interest is the *security patrol problem*, which was suggested by Maria Fox as possibly relevant to the theory of word-representable graphs. Here, a security guard patrols a network of connected locations, and there might be constraints determining the order in which some of the locations are visited and preventing locations from being revisited before other parts of the network have been patrolled. Problems like this are of great importance in a range of manufacturing, rostering, surveillance and defence applications, so finding ways to exactly characterize the key constraints, and thereby eliminate unnecessary search amongst partial solutions that do not satisfy them, would have great potential impact. In addition, the non-existence of words representing the graph of such a problem will identify the boundary between solvable and unsolvable instances, which could be of great value in solving planning and scheduling problems.

Chapter 2

Hereditary Classes of Graphs

The main subject of this book, the class of word-representable graphs, belongs to a wide family of graph classes known as *hereditary*. In this chapter, we will take a little tour through the world of hereditary classes. In particular, we will survey basic results related to this notion and will introduce various hereditary classes pertinent to word-representable graphs. We start with a formal definition.

Definition 2.0.1. A class X of graphs containing with each graph G all induced subgraphs of G is called *hereditary*. In other words, X is hereditary if and only if $G \in X$ implies $G - v \in X$ for any vertex v of G, where $G - v$ denotes the graph obtained from G by removing v together with all edges incident to it.

The family of hereditary classes contains many classes of theoretical or practical importance, such as bipartite graphs, permutation graphs, comparability graphs etc. Many classes that are not hereditary have natural hereditary extensions. For instance, the class of trees can be extended to the hereditary class of forests, and the class of k-regular graphs to the hereditary class of graphs of vertex degree at most k. These and various other hereditary classes relevant to the topic of the book will be defined in Section 2.2.

Hereditary classes enjoy many nice properties. Some of them are listed in the next lemma, which is easy to see.

Lemma 2.0.2. *Let X and Y be hereditary classes of graphs, and let $\overline{X}$ denote the class of complements of graphs in X. Then classes $X \cup Y$, $X \cap Y$ and $\overline{X}$ are also hereditary.*

One more important property of hereditary classes is that they admit so-called *forbidden induced subgraph characterization.* We define this notion and present several results related to it in the next section.

2.1 Induced Subgraph Characterization of Hereditary Classes

Given a set of graphs M (finite or infinite), we denote by

> $Free(M)$ the class of graphs containing no graph from M as an induced subgraph.

We say that graphs in M are *forbidden induced subgraphs* for the class $Free(M)$ and that graphs in $Free(M)$ are M-free.

Theorem 2.1.1. *A class X of graphs is hereditary if and only if there is a set M such that $X = Free(M)$.*

Proof. Assume $X = Free(M)$ for a set M. Consider a graph $G \in X$ and an induced subgraph H of G. Then H is M-free, since otherwise G contains a forbidden graph from M. Therefore, $H \in X$ and hence X is hereditary.

Conversely, if X is hereditary, then $X = Free(M)$ with M being the set of all graphs not in X. □

To illustrate the theorem, consider the set X of all complete graphs. Clearly, this set is hereditary and $X = Free(M)$ with M being the set of all non-complete graphs. On the other hand, it is not difficult to see that $X = Free(\overline{K}_2)$, since a graph G is complete if and only if $\overline{K}_2$ is not an induced subgraph of G, i.e. G has no pair of non-adjacent vertices. This example motivates the following definition.

Definition 2.1.2. A graph G is a *minimal* forbidden induced subgraph for a hereditary class X if G does not belong to X but every proper induced subgraph of G belongs to X (or alternatively, the deletion of any vertex from G results in a graph that belongs to X).

Let us denote by

> $Forb(X)$ the set of all minimal forbidden induced subgraphs for a hereditary class X.

Theorem 2.1.3. *For any hereditary class X, we have $X = Free(Forb(X))$. Moreover, $Forb(X)$ is the unique minimal set with this property.*

Proof. To prove that $X = Free(Forb(X))$, we show two inclusions:

$$X \subseteq Free(Forb(X)) \text{ and } Free(Forb(X)) \subseteq X.$$

Assume first $G \in X$, then by definition all induced subgraphs of G belong to X and hence no graph from $Forb(X)$ is an induced subgraph of G, since none of them belongs to X. As a result, $G \in Free(Forb(X))$, which proves that $X \subseteq Free(Forb(X))$.

Assume now that $G \in Free(Forb(X))$, and suppose by contradiction that $G \notin X$. Let H be a minimal induced subgraph of G which is not in X (possibly $H = G$). But then $H \in Forb(X)$ contradicting the fact that $G \in Free(Forb(X))$. This contradiction shows that $G \in X$ and hence proves that $Free(Forb(X)) \subseteq X$.

To prove the minimality of the set $Forb(X)$, we will show that for any set N such that $X = Free(N)$ we have $Forb(X) \subseteq N$. Assume this is not true and let H be a graph in $Forb(X) - N$. By the minimality of the graph H, any proper induced subgraph of H is in X, and hence is in $Free(N)$. Together with the fact that H does not belong to N, we conclude that $H \in Free(N)$. Therefore $H \in Free(Forb(X))$. But this contradicts the fact that $H \in Forb(X)$. □

The induced subgraph characterization of a hereditary class is important for many reasons. For instance, it provides an easy way to decide whether a hereditary class is finite or not, which is due to the following theorem.

Theorem 2.1.4. *A hereditary class $X = Free(M)$ is finite if and only if M contains a complete graph and an edgeless graph.*

Proof. If M contains no complete graph, then all complete graphs belong to $X = Free(M)$ and hence X is infinite. Similarly, X is infinite if M contains no edgeless graph.

If both a complete graph K_n and an edgeless graph O_m are forbidden for X, then graphs in X have at most $R(n,m)$ vertices, where $R(n,m)$ is the Ramsey number, and hence X is a finite class. □

Detour.

To better illustrate the importance of the induced subgraph characterization of hereditary classes, let us consider the following example. In 1969, the "Journal of Combinatorial Theory" published a paper entitled "An interval graph is a comparability graph" [82]. One year later, the same journal published another paper entitled "An interval graph is not a comparability graph" [60]. With the induced subgraph characterization this situation could not happen, because of the following theorem.

Theorem 2.1.5. *$Free(M_1) \subseteq Free(M_2)$ if and only if for every graph $G \in M_2$ there is a graph $H \in M_1$ such that H is an induced subgraph of G.*

Proof. Assume $Free(M_1) \subseteq Free(M_2)$, and suppose to the contrary that a graph $G \in M_2$ contains no induced subgraphs from M_1. By definition, this means that $G \in Free(M_1) \subseteq Free(M_2)$. On the other hand, G belongs to the set of forbidden graphs for $Free(M_2)$, a contradiction.

Conversely, assume that every graph in M_2 contains an induced subgraph from M_1. By contradiction, let $G \in Free(M_1) - Free(M_2)$. Since G does not belong to $Free(M_2)$, by definition G contains an induced subgraph $H \in M_2$. Due to the assumption, H contains an induced subgraph $H' \in M_1$. Then obviously H' is an induced subgraph of G which contradicts the fact that $G \in Free(M_1)$. □

Therefore, given two hereditary classes of graphs and the induced subgraph characterization for both of them, it is a simple task to decide the inclusion relationship between them. Both the interval graphs and the comparability graphs form hereditary classes (to be discussed later). Apparently, in 1969 the induced subgraph characterization was not available for at least one of these two classes. Nowadays, it is available for both of them.

This example shows that the problem of finding the set of minimal forbidden induced subgraphs for a hereditary class X is an important task. However, this problem is in general far from being trivial. For instance, for the class of *perfect graphs* this problem was open for several decades [36]. The class of perfect graphs and several related classes are introduced in the next section.

We conclude the present section with two simple properties of hereditary classes related to sets of forbidden induced subgraphs.

Lemma 2.1.6. *Let $X = Free(M_1)$ and $Y = Free(M_2)$ be hereditary classes and for an arbitrary set Z of graphs, let $\overline{Z}$ denote the set of complements of graphs in Z. Then $X \cap Y = Free(M_1 \cup M_2)$ and $\overline{X} = Free(\overline{M}_1)$.*

The proof of this lemma is left to the reader as an exercise (see Section 2.6).

2.2 Examples of Hereditary Classes

The world of hereditary classes is rich and contains many classes of fundamental importance. In this section, we introduce several hereditary classes relevant to the topic of this book. We start with the description of classes defined by small forbidden induced subgraphs.

As we have seen already, $Free(\overline{K}_2)$ is the class of complete graphs. Similarly, $Free(K_2)$ is the class of edgeless graphs. Now let us consider some classes defined by forbidden induced subgraphs with three vertices. Up to isomorphism, there are four such graphs: $P_3, \overline{P}_3, K_3, \overline{K}_3$. The structure of P_3-free graphs and their complements is very simple and can be characterized as follows.

Lemma 2.2.1. *A graph G is P_3-free if and only if it is a disjoint union of cliques, i.e. every connected component of G is a clique.*

Proof. Clearly, if G is a disjoint union of cliques, then G is P_3-free . Conversely, let G be a P_3-free graph and assume it contains a connected component which is not a clique. Consider any two non-adjacent vertices in this component and a shortest (i.e. chordless) path connecting them. This path contains at least two edges, and therefore, G contains a P_3. This contradiction completes the proof. □

From Lemma 2.2.1 it follows that the complement of a P_3-free graph is a graph whose vertices can be partitioned into independent sets so that any two vertices belonging to different independent sets are adjacent. Such graphs are called *complete multipartite.*

Corollary 2.2.2. *A graph is complete multipartite if and only if it is $\overline{P}_3$-free.*

The class of K_3-free graphs (also known as triangle-free graphs) is much richer and contains, for instance, all bipartite graphs.

2.2.1 Bipartite Graphs

Definition 2.2.3. A graph is *bipartite* if its vertices can be partitioned into at most two independent sets.

From the definition it follows that the class of bipartite graphs is hereditary and hence it can be characterized in terms of minimal forbidden induced subgraphs. It is easy to see that cycles C_n of odd length are (minimal) non-bipartite graphs, and therefore the bipartite graphs form a subclass of $Free(C_3, C_5, C_7, \ldots)$. The inverse inclusion was proved by König [99]. Thus the bipartite graphs can be characterized as follows.

Theorem 2.2.4. *The class of bipartite graphs is precisely the class*

$$Free(C_3, C_5, C_7, \ldots).$$

A bipartite graph $G = (V, E)$ given together with a bipartition $V = A \cup B$ is denoted $G = (A, B, E)$. In the case when every vertex of A is adjacent to every vertex of B, the graph G is called *complete bipartite*. Allowing one of the parts in the bipartition to be empty, we conclude that the class of complete bipartite graphs is hereditary. In terms of forbidden induced subgraphs this class can be characterized as follows.

Theorem 2.2.5. *A graph is complete bipartite if and only if it is* $(\overline{P}_3, K_3)$*-free.*

Proof. Neither $\overline{P}_3$ nor K_3 is a complete bipartite graph, therefore a complete bipartite graph is $(\overline{P}_3, K_3)$-free. Conversely, if a graph is $\overline{P}_3$-free, then it is complete multipartite, and if it is K_3-free, then the number of parts cannot be larger than two, since otherwise a K_3 arises. □

An important subclass of bipartite graphs is the class of forests.

Definition 2.2.6. A *forest* is a graph without cycles, also known as an acyclic graph.

In other words, a graph G is a forest if and only if every connected component of G is a tree. An important generalization of forests is the class of *chordal graphs*, also known as *triangulated graphs.*

2.2.2 Chordal Graphs, Split Graphs and Interval Graphs

Definition 2.2.7. A graph is *chordal* if and only if it is $(C_4, C_5, C_6, \ldots)$-free. Chordal graphs are also known as *triangulated graphs.*

By Lemma 2.1.6 the class $Free(\overline{C}_4, \overline{C}_5, \overline{C}_6, \ldots)$ contains complements of chordal graphs, and by the same lemma $Free(C_4, \overline{C}_4, C_5, \overline{C}_5, C_6, \overline{C}_6, \ldots)$ is the intersection of the two classes, chordal graphs and their complements. However, the list of forbidden graphs describing this intersection is redundant (i.e. not all graphs in the list are minimal). Indeed, it is not difficult to see that the complement of C_4 is $2K_2$ (i.e. the disjoint union of two K_2s), and every cycle of length at least 6 contains a $2K_2$ as an induced subgraph. Similarly, the complement of every cycle of length at least 6 contains a C_4 as an induced subgraph. Finally, C_5 is self-complementary, and therefore, the intersection of the classes of chordal graphs and their complements can be characterized by three minimal forbidden induced subgraphs: $C_4, 2K_2$ and C_5 .

Graphs which are $(C_4, 2K_2, C_5)$-free possess a nice decomposability property proved in [61].

Lemma 2.2.8. *([61]) The vertices of every $(C_4, 2K_2, C_5)$-free graph can be partitioned into a clique and an independent set.*

The set of all graphs possessing the decomposability property of Lemma 2.2.8 are known as *split graphs.*

Definition 2.2.9. A graph is a *split graph* if and only if its vertices can be partitioned into a clique and an independent set.

Thus Lemma 2.2.8 shows that every $(C_4, 2K_2, C_5)$-free graph is a split graph. On the other hand, it is not difficult to see that none of the graphs C_4, $2K_2$ and C_5 is a split graph, and moreover, each of them is a minimal non-split graph. Summarizing the above discussion we reach the following conclusion.

Theorem 2.2.10. *The class of split graphs is the intersection of the classes of chordal graphs and their complements, and this is precisely the class $Free(C_4, 2K_2, C_5)$.*

One more important subclass of chordal graphs is known as the *interval graphs.*

Definition 2.2.11. A graph is an *interval graph* if it is the intersection graph of a family of intervals on a straight line, i.e. the graph whose vertices are the intervals and the edges are pairs of intervals with a non-empty intersection.

Directly from the definition we conclude that the class of interval graphs is hereditary. Also, it is not difficult to see that chordless cycles of length 4 or more are not interval graphs, i.e. every interval graph is chordal. Moreover, the cycles $C_4, C_5, C_6, \ldots$ are *minimal* non-interval graphs. However, this list of minimal non-interval graphs is not complete, and the complete list was found in [103].

2.2.3 Comparability Graphs and Permutation Graphs

Definition 2.2.12. A graph G is a *comparability* (also known as *transitively orientable*) graph if and only if the edges of G can be oriented so that the obtained orientation is transitive, i.e. an arc from x to y and an arc from y to z imply an arc from x to z.

It is not difficult to see that the deletion of a vertex from a transitively oriented graph cannot destroy the transitivity. Therefore, the class of comparability graphs is hereditary and hence can be characterized by a set of minimal forbidden induced subgraphs.

It is a simple exercise to verify that odd cycles of length at least 5 are minimal non-comparability graphs. However, the list of minimal non-comparability graphs is much more complicated and contains many more graphs. This list was found by Gallai in [63].

The class of comparability graphs contains several important subclasses. For instance, all bipartite graphs are transitively orientable. Indeed, let $G = (A, B, E)$ be a bipartite graph given together with a bipartition of its vertices into two independent sets A and B. Then by orienting all edges of G from A to B we conclude that G is transitively orientable.

A less obvious inclusion, proved by Gilmore and Hoffman in [66], states that the complements of interval graphs are comparability graphs. Moreover, Gilmore and Hoffman [66] proved that the class of interval graphs is precisely the intersection of chordal graphs and co-comparability graphs (i.e. complements of comparability graphs).

To introduce one more important subclass of comparability graphs, let us consider a permutation π of the set $\{1, 2, \dots, n\}$. A pair (i, j) is called an *inversion* in π if $(i - j)(\pi(i) - \pi(j)) < 0$. For instance, the pair $(2, 5)$ is an inversion in the permutation 15342.

Definition 2.2.13. The graph of a permutation π, denoted $G[\pi]$, has $\{1, 2, \dots, n\}$ as its vertex set with i and j being adjacent if and only if (i, j) is an inversion.

One more way of defining the graph of a permutation π is by representing the permutation in the form of a diagram as illustrated on the left of Figure 2.1. Then $G[\pi]$ is the intersection graph of the diagram representing π, i.e. the graph whose vertices are the line segments with two vertices being adjacent if and only if the respective segments intersect (cross) each other (again, see Figure 2.1 for an illustration).

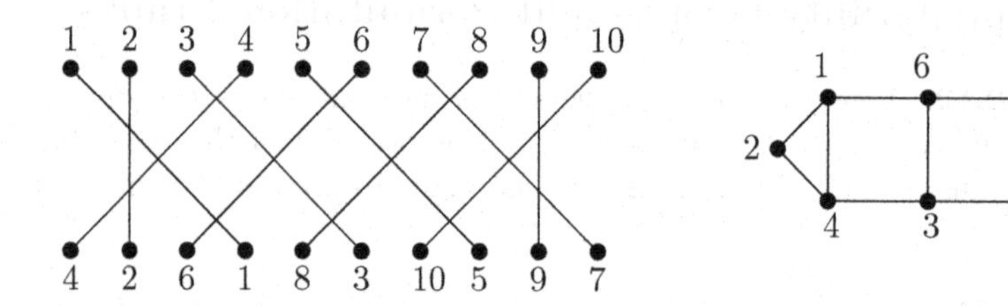

Figure 2.1: The diagram representing the permutation $\pi = 426183(10)597$ (left) and the permutation graph G_π (right)

Definition 2.2.14. A graph G is a *permutation graph* if there is a permutation π such that G is isomorphic to $G[\pi]$.

Obviously, the deletion of a line segment from the diagram representing a permutation results in a diagram representing another permutation. Therefore, the deletion of a vertex from a permutation graph results in a permutation graph again. Thus, the class of permutation graphs is hereditary.

Let G be the permutation graph of a permutation π. By orienting every edge of G from a smaller element of π to a larger element, we conclude that this orientation is transitive. Therefore, we have the following theorem.

Theorem 2.2.15. *Every permutation graph is a comparability graph.*

Also, by reversing the order of the bottom endpoints of the line segments in the diagram representing π, we obtain a diagram of another permutation on the same set, say π'. This operation transforms crossing segments into non-crossing ones, and vice versa. Therefore, $G[\pi']$ is the complement of $G[\pi]$ and hence we have the following result.

Theorem 2.2.16. *The complement of a permutation graph is a permutation graph.*

From the two previous theorems we conclude that permutation graphs belong to the intersection of comparability graphs and their complements. A more intriguing and deeper result, proved in [49], states that the class of permutation graphs coincides with this intersection.

Theorem 2.2.17. *([49]) The class of permutation graphs is the intersection of the class of comparability graphs and the class of their complements.*

An important generalization of permutation graphs is the class of *circle graphs* defined in the next section.

2.2.4 Circle Graphs and Planar Graphs

Definition 2.2.18. A *circle graph* is the intersection graph of a set of chords of a circle.

An example of a circle graph can be found in Figure 5.2.

Since the deletion of a vertex from a circle graph is equivalent to the deletion of a chord from a circle, the class of circle graphs is clearly hereditary and therefore there must exist a characterization of this class in terms of minimal forbidden induced subgraphs. Unfortunately, the complete list of these graphs is still unavailable. However, this class admits a nice characterization in terms of other types of obstructions. To explain this idea, we need to make a small detour.

Detour: *partial orders on graphs.*

We repeat that a class of graphs is hereditary if it is closed under taking induced subgraphs and that a graph G is an induced subgraph of a graph H if G can be obtained from H by a (possibly empty) sequence of vertex deletions. However, the induced subgraph relation is not the only partial order on graphs. There are many other partial orders studied in the literature, such as *subgraphs* (in which case in addition to vertex deletions we are allowed to delete edges), *minors* (vertex deletions, edge deletions and edge contractions), *induced minors* (vertex deletions and edge contractions) etc.

Many classes of graphs that are closed under taking induced subgraphs (i.e. hereditary classes) are also closed under other partial orders (subgraphs, minors etc.), in which case such classes can be characterized in terms of other obstructions (minimal forbidden subgraphs, minors etc.). For instance, it is not difficult to see that the class of bipartite graphs is closed under taking subgraphs (not necessarily induced) and hence it can be characterized in terms of minimal forbidden subgraphs. However, in this case the subgraph characterization and induced subgraph characterization coincide, because it is easy to see that a graph contains an odd cycle as a subgraph if and only if it contains an odd cycle as an induced subgraph.

A more interesting example is the class of interval graphs, which, in addition to being closed under taking induced subgraphs, is also closed under taking induced minors. Indeed, the contraction of an edge uv of an interval graph is equivalent to the union of the intervals representing vertices u and v. Therefore, a graph obtained by contracting an edge of an interval graph is again interval. In [57], the class of interval graphs has been characterized in terms of five minimal forbidden induced minors.

Perhaps the most famous example of a hereditary class characterized by means of obstructions different from forbidden induced subgraphs is the class of planar graphs.

Definition 2.2.19. A graph is *planar* if it can be drawn on the plane in such a way that no two edges cross each other.

It is well known (and not difficult to see) that the class of planar graphs is closed under taking minors, i.e. vertex deletion, edge deletion or edge contraction applied to a planar graph results in a planar graph. A famous theorem of Kuratowski characterizes the class of planar graphs in terms of minimal forbidden minors.

Theorem 2.2.20. *([102]) A graph is planar if and only if it contains neither K_5 nor $K_{3,3}$ as a minor.*

To find a description of the class of circle graphs alternative to the induced subgraph characterization, let us observe that this class is closed under *local complementation*, i.e. the operation of complementing the edges in the neighbourhood of a vertex. Indeed, the local complementation applied to the neighbourhood of a vertex v in a circle graph G can be performed on the circle diagram representing G as follows: we cut the circle along the chord representing v and then twist half of the circle along the cut. This explanation shows that the graph obtained from a circle graph by local complementation is again a circle graph.

A graph obtained from a graph G by a (possibly empty) sequence of vertex deletions and local complementations is called a *vertex-minor* of G. Thus the class of circle graphs is closed under the vertex minor relation. In [26], this class was characterized by three minimal forbidden vertex-minors.

Theorem 2.2.21. *([26]) A graph is a circle graph if and only if it does not contain the three graphs in Figure 2.2 as vertex-minors.*

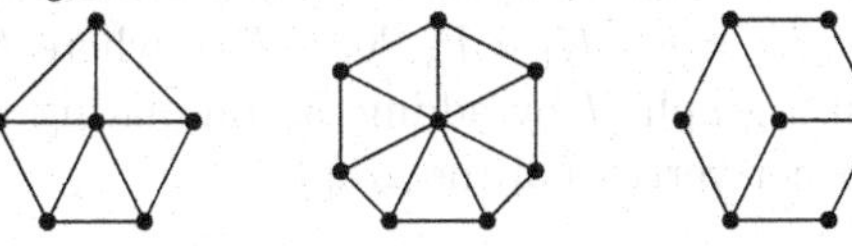

Figure 2.2: Three minimal forbidden vertex-minors for the class of circle graphs

We have mentioned already that the class of circle graphs extends the class of permutation graphs. To see this, simply encircle the intersection diagram representing a permutation. One more important subclass of circle graphs is the class of *outerplanar graphs*, i.e. planar graphs that can be drawn in such a way that all of the vertices belong to the unbounded face of the drawing. A formal definition and examples of outerplanar graphs can be found in Definition 5.1.4 and Example 5.1.5, respectively.

We conclude this section with an intriguing relationship between circle bipartite graphs and planar graphs revealed in [46], where we refer to that paper for definitions of a *fundamental graph* and *planar matroid*.

Theorem 2.2.22. *([46]) A graph is a circle bipartite graph if and only if it is a fundamental graph of a planar matroid.*

2.2.5 k-Colourable Graphs and Graphs of Vertex Degree at most k

The *degree* of a vertex v is the number of its neighbours, or equivalently, the number of edges incident to v. For any fixed $k \geq 0$, the set of all graphs with maximum vertex degree at most k forms a hereditary class, since the deletion of a vertex from a graph in this class does not increase the degree of the remaining vertices.

Let D_k denote the class of graphs of degree at most k and let G be a graph in D_k.

- If $k = 0$, then G is an edgeless graph and $D_0 = Free(K_2)$.
- If $k = 1$, then G is a graph every connected component of which has at most two vertices and $D_1 = Free(P_3, K_3)$. Indeed, $D_1 \subseteq Free(P_3, K_3)$, since P_3 and K_3 have vertices of degree 2. On the other hand, if a graph is P_3-free, then every connected component of this graph is a clique by Lemma 2.2.1, and if in addition it is K_3-free, then every component has size at most 2, i.e. $Free(P_3, K_3) \subseteq D_1$.
- If $k = 2$, then every connected component of G is either a path or a cycle and $D_2 = Free(K_1 \times K_3, K_1 \times \overline{K}_3, K_1 \times P_3, K_1 \times \overline{P}_3)$, where $K_1 \times H$ denotes the graph obtained from a graph H by adding a dominating vertex, i.e. a vertex adjacent to all the other vertices of the graph.

In general, it is not difficult to see that

$$D_k = Free(\{K_1 \times H \ : \ H \text{ is a graph with } k+1 \text{ vertices}\}).$$

Definition 2.2.23. A graph G is *k-colourable* if the vertex set of G can be partitioned into at most k independent sets. The minimum value of k such that G is k-colourable is the *chromatic number* of G, denoted $\chi(G)$.

The celebrated Brooks' Theorem establishes an interesting relationship between the maximum vertex degree in a graph and its chromatic number.

Theorem 2.2.24. *(Brooks' Theorem, [29]) For any connected graph G with maximum vertex degree Δ, the chromatic number of G is at most Δ, unless G is a complete graph or an odd cycle, in which case the chromatic number is $\Delta + 1$.*

Corollary 2.2.25. *Every connected graph of vertex degree at most 3 is 3-colourable, except for K_4.*

Clearly, for each fixed value of $k \geq 1$, the set of all k-colourable graphs, or equivalently graphs of chromatic number at most k, constitute a hereditary class. However, the characterization of k-colourable graphs in terms of minimal forbidden induced subgraphs is known only for $k = 1$ and $k = 2$. Specifically, for $k = 1$, this class coincides with the set of all edgeless graphs, i.e. with the class

$Free(K_2)$, and for $k = 2$, this is precisely the class of bipartite graphs, i.e. the class $Free(C_3, C_5, C_7, \ldots)$. For larger values of k, only partial information is available. In particular, for each value of k, the complete graph on $k + 1$ vertices is a minimal non-k-colourable graph. This follows from the obvious fact (and also from Brooks' Theorem) that the chromatic number of a complete graph equals the number of its vertices. This fact also shows that the chromatic number of a graph can never be smaller than its clique number, i.e. $\chi(G) \geq \omega(G)$ for any graph G. A hereditary class of graphs for which this inequality becomes equality is of particular interest and it is defined in the next section.

2.2.6 Perfect Graphs and Related Classes

Definition 2.2.26. A graph G is *perfect* if $\chi(H) = \omega(H)$ for every induced subgraph H of G.

By definition, the class of perfect graphs is hereditary and therefore there must exist a characterization of this class in terms of minimal forbidden induced subgraphs. This characterization was obtained in [36] and is known as the *Strong Perfect Graph Theorem.* However, the result obtained in [36] was conjectured much earlier and its proof was preceded by numerous other contributions supporting the conjecture.

Detour: *Chasing perfection.*

In the late 1950s to early 1960s, Claude Berge studied several classes of graphs one of which is the class of perfect graphs (as defined in Definition 2.2.26) and one is the class of their complements. He conjectured that the two classes coincide (*Berge's Weak Perfect Graph Conjecture*) and that both are precisely the graphs containing neither odd cycles of length at least 5 nor their complements (*Berge's Strong Perfect Graph Conjecture*). The weak conjecture was proved by Lovász in 1972 [105] and is known nowadays as the *Perfect Graph Theorem.* However the strong conjecture of Berge was around for more than 40 years and attracted the attention of hundreds of researchers.

It is a simple exercise to verify that odd cycles of length at least 5 (also known as odd holes) are minimal imperfect graphs. Indeed, the chromatic number of any odd hole is 3, while its clique number is 2, and hence odd holes are imperfect. Also, by deleting any vertex from an odd hole we obtain a path, which is obviously a perfect graph, and hence odd holes are minimal imperfect graphs. A more tricky conclusion (which follows, in particular, from the Perfect Graph Theorem) is that the complements of odd holes (also known as odd antiholes) are also minimal imperfect graphs. Whether there are more minimal imperfect graphs is the essence of the Strong Perfect Graph Conjecture. In the attempt to answer this question dozens of properties of minimal imperfect graphs were discovered in the literature, and the conjecture was verified for many restricted instances (for *claw-free graphs*, *bull-free graphs* etc). A complete proof was announced in 2002 by a team of four people: M. Chudnovsky, N. Robertson, P. Seymour and R. Thomas. The proof was published in 2006 in the Annals of Mathematics [36].

The importance of perfect graphs is due to the fact that they link various mathematical disciplines, such as graph theory, combinatorial optimization, semidefinite programming, polyhedral and convexity theory, and even information theory. Also, they contain many other important graph families as subclasses. In fact, most of the graph classes mentioned before are perfect. This is the case for bipartite graphs, comparability graphs, chordal graphs, permutation graphs, interval graphs and their complements.

The theory of perfect graphs also motivated the study of various related concepts. For instance, the paper [112] studies *edge-perfectness*. Another important notion motivated by perfect graphs is *χ-boundedness*. A hereditary class of graphs is said to be χ-bounded if for every graph G in the class the chromatic number of G is bounded by a function of its clique number. Clearly, perfectness is just a special case of χ-boundedness.

2.2.7 Line Graphs

Definition 2.2.27. The *line graph* of a graph $G = (V, E)$ is the graph with vertex set E in which two vertices are adjacent if and only if the corresponding edges of G share a vertex. The line graph of G is denoted $L(G)$.

Let $G = (V, E)$ be a graph and $e \in E$. Then clearly $L(G) - e$ is the line graph of $G - e$. Therefore, the class of line graphs is hereditary. It is not difficult to verify that $K_{1,3}$, the claw, is not a line graph, i.e. there is no graph G such that $L(G) = K_{1,3}$. Moreover, $K_{1,3}$ is a minimal non-line graph, i.e. the deletion of any vertex from $K_{1,3}$ results in a line graph. The complete list of minimal forbidden induced subgraphs for the class of line graphs consists of nine graphs and can be found in [77].

From the definition it follows that a matching in a graph G corresponds to an independent set in $L(G)$. Since the MAXIMUM MATCHING problem is solvable in polynomial time, so is the MAXIMUM INDEPENDENT SET problem when restricted to the class of line graphs. We observe that for general graphs the maximum independent set problem is NP-complete.

According to König's Theorem, in a bipartite graph G the size of a maximum matching equals the size of a minimum vertex cover. This implies that in the complement of the line graph of G, the clique number equals the chromatic number. Therefore, both $L(G)$ and its complement are perfect graphs, when G is perfect. Moreover, the line graphs of bipartite graphs and their complements are two of the five basic building blocks in the proof of the Strong Perfect Graph Theorem.

2.3 On the Size of Hereditary Graph Classes

In this section, we discuss several results on the number of graphs in hereditary classes. Given a class X, we write X_n for the set of graphs in X with vertex set $\{1, 2, \ldots, n\}$, i.e. X_n is the set of *labelled* graphs in X. Following [16], we call $|X_n|$ the *speed* of X.

Clearly, the number of all labelled graphs on n vertices is $2^{\binom{n}{2}}$, since each of the $\binom{n}{2}$ pairs of vertices can be either adjacent or not. Computing the speed for specific graph classes is generally a more difficult task. However, for some classes with simple structure, one can still produce a computable formula for $|X_n|$. For instance,

- If X is the class of complete graphs, then $|X_n| = 1$ for each value of n.
- If $X = Free(P_3, 2K_2)$, then $|X_n| = 2^n - n$. Indeed, by Lemma 2.2.1 every connected component of a P_3-free graph G is a clique and due to the $2K_2$-freeness at most one connected component has more than one vertex. Therefore, every graph in $Free(P_3, 2K_2)$ can be partitioned into a clique and a set of isolated vertices. It is not difficult to see that the number of labelled graphs admitting such a partition is $2^n - n$.
- If $X = Free(P_3, K_3)$, then X is the class of graphs of vertex degree at most 1 (see Section 2.2.5), or equivalently, the class of graphs every connected component of which has at most two vertices. In this case $|X_n|$ coincides with the number of involutions (permutations consisting of cycles of length at most 2) and hence can be computed as follows:

$$|X_n| = \sum_{k=0}^{\lfloor n/2 \rfloor} \frac{n!}{(n-2k)!k!2^k}.$$

- If $X = Free(P_3)$, then $|X_n|$ is the nth *Bell number*, i.e. the number of partitions of an n-member set into non-empty subsets, and hence can be computed as follows:

$$|X_n| = \sum_{k=0}^{n} \frac{1}{k!} \sum_{j=0}^{k} (-1)^{k-j} \binom{k}{j} j^n.$$

However, in most cases exact formulas for $|X_n|$ are not available, in which case we restrict ourselves to approximate or asymptotic formulas or bounds on $|X_n|$. Of special interest is the asymptotic behaviour of the following ratio:

$$\frac{\log_2 |X_n|}{\binom{n}{2}}.$$

To explain the importance of this ratio, let us observe that there is a close relationship between counting graphs and their coding (i.e. representation in a finite alphabet). A classical example exhibiting this relationship is the *Prüfer code* for labelled trees [120] considered in Section 8.1.1. It assigns to each tree with n vertices a word of length $n-2$ in an alphabet of n letters in such a way that there is a bijection between the words and the trees. This proves that the number of labelled trees with n vertices is n^{n-2}, which is known as *Cayley's formula.*

Clearly, every graph with n vertices can be represented by a binary word of length $\binom{n}{2}$ (one bit per pair of vertices). If we want to represent graphs in a particular class X, we may try to reduce the length of the representation and $\log_2 |X_n|$ indicates the minimum length of binary words representing graphs in X_n, since the number of words cannot be smaller than the number of graphs. Therefore, the ratio

$$\frac{\log_2 |X_n|}{\binom{n}{2}}$$

can be viewed as the coefficient of compressibility for representing n-vertex graphs in X. Its limit value, for $n \to \infty$, was called by Alekseev in [5] the *entropy* of X. Moreover, in the same paper Alekseev showed that for every hereditary property X the entropy necessarily exists and in [6] he proved that its value has the following form:

$$\lim_{n\to\infty} \frac{\log_2 |X_n|}{\binom{n}{2}} = 1 - \frac{1}{c(X)}, \tag{2.1}$$

where $c(X)$ is a natural number, called the *index* of X. To define this notion let us denote by $\mathcal{E}_{i,j}$ the class of graphs whose vertices can be partitioned into at most i independent sets and j cliques. In particular, $\mathcal{E}_{2,0}$ is the class of bipartite graphs and $\mathcal{E}_{1,1}$ is the class of split graphs. Then $c(X)$ is the largest k such that X contains $\mathcal{E}_{i,j}$ with $i+j=k$. Independently, a similar result was obtained by Bollobás and Thomason [24, 25] and is known nowadays as the Alekseev-Bollobás-Thomason Theorem (see e.g. [9]).

2.4 Coding of Graphs in Hereditary Classes

By graph coding we mean the problem of representing graphs by words in a finite alphabet. This problem is important in computer science for representing graphs in computer memory [64, 85, 129]. Without loss of generality we will assume that our alphabet is binary, i.e. consists of two symbols 0 and 1.

More formally, denote $B = \{0,1\}$ and let B^* be the set of all words in the alphabet B. Given a word $\alpha \in B^*$, we denote by $|\alpha|$ the length of α and by α_j the jth letter of α. Also, let λ represent the empty word (the only word of length 0).

A *coding* of graphs in a class X is a family of bijective mappings $\Phi = \{\phi_n : n = 1,2,3,\ldots\}$, where $\phi_n : X \to B^*$. A coding Φ is called *asymptotically optimal* if

$$\lim_{n\to\infty} \frac{\max\limits_{G\in X_n} |\phi_n(G)|}{\log_2 |X_n|} = 1.$$

As we mentioned earlier, every labelled graph G with n vertices can be represented by a binary word of length $\binom{n}{2}$, one bit per pair of vertices, with 1 standing for an edge and 0 for an non-edge. Such a word can be obtained by reading the elements of the adjacency matrix above the main diagonal. The word obtained by reading these elements row by row, starting with the first row, will be called the *canonical coding* of G and will be denoted $\phi_n^c(G)$.

If no prior information about the graph is available, then $\binom{n}{2}$ is the minimum number of bits needed to represent the graph. However, if we know that our graph possesses some special properties, then this knowledge may lead to a shorter representation. For instance,

- If we know that our graph is bipartite, then we do not need to describe the adjacency of vertices that belong to the same part in its bipartition. Therefore, we need at most $n^2/4$ bits to describe the graph, the worst case being a bipartite graph with $n/2$ vertices in each of its parts.

- If we know that our graph is not an arbitrary bipartite graph but *chordal* bipartite, then we can further shorten the code and describe any graph in this class with at most $O(n \log^2 n)$ bits [128].

- A further restriction to trees (a proper subclass of chordal bipartite graphs) enables us to further shorten the code to $(n-2)\log n$ bits, which is the length of the binary representation of the Prüfer code for trees [120].

The notion of entropy defined in the previous section suggests how much the canonical representation can be shortened for graphs in a hereditary class. In what follows we discuss the question of how to achieve an optimal representation. For hereditary classes of index $c > 1$, i.e. of non-zero entropy, an answer to this question was proposed in [5] by Alekseev, who described a universal algorithm which gives an asymptotically optimal coding for graphs in every hereditary class X of index $c > 1$. Below we present an adapted version of this algorithm with a proof of its optimality.

Let $n > 1$ and p be a prime number between $\lfloor n/\sqrt{\log_2 n}+1\rfloor$ and $2\lfloor n/\sqrt{\log_2 n}\rfloor$. Such a number always exists by the Bertrand-Chebyshev Theorem (see e.g. [2]). Define $k = \lfloor n/p\rfloor$. Then

$$p \le 2n/\sqrt{\log_2 n}, \quad k \le \sqrt{\log_2 n}, \quad n - kp < p. \tag{2.2}$$

Let G be an arbitrary graph with n vertices. We split the set of all pairs of vertices of G into two disjoint subsets R_1 and R_2 as follows: R_1 consists of the pairs (a, b) such that $a \le kp$, $b \le kp$ and $\lfloor (a-1)/p\rfloor \ne \lfloor (b-1)/p\rfloor$, and R_2 consists of all the remaining pairs. Let us denote by $\mu^{(1)}$ the binary word consisting of the elements of the canonical code corresponding to the pairs of R_2. This word will be included in the code of G we construct.

Now let us examine the pairs in R_1. For all $x, y \in \{0, 1, \ldots, p-1\}$, we define

$$Q_{x,y} = \{pi + 1 + res_p(xi + y) \ : \ i = 0, 1, \ldots, k-1\},$$

where $res_p(z)$ is the remainder on dividing z by p. Let us show that every pair of R_1 appears in exactly one set $Q_{x,y}$. Indeed, if $(a, b) \in Q_{x,y}$ $(a < b)$, then

$$xi_1 + y \equiv a \pmod{p}, \quad xi_2 + y \equiv b \pmod{p},$$

where $i_1 = \lfloor (a-1)/p\rfloor$, $i_2 = \lfloor (b-1)/p\rfloor$. Since $i_1 \ne i_2$ (by definition of R_1), there exists a unique solution of the following system

$$\begin{aligned} x(i_1 - i_2) &\equiv a - b \pmod{p} \\ y(i_1 - i_2) &\equiv ai_2 - bi_1 \pmod{p}. \end{aligned} \tag{2.3}$$

Therefore, by coding the graphs $G_{x,y} = G[Q_{x,y}]$ and combining their codes with the word $\mu^{(1)}$ (which describes the pairs in R_2) we obtain a complete description of G.

To describe the graphs $G_{x,y} = G[Q_{x,y}]$ we first relabel their vertices according to

$$z \to \lfloor (z-1)/p\rfloor + 1.$$

In this way, we obtain p^2 graphs $G'_{x,y}$, each on the vertex set $\{1, 2, \ldots, k\}$. Some of these graphs may coincide. Let m $(m \le p^2)$ denote the number of pairwise different graphs in this set and $(H_0, H_1, \ldots, H_{m-1})$ an ordered list of m pairwise different graphs in this set. In other words, for each graph $G'_{x,y}$ there is a unique number i such that $G'_{x,y} = H_i$. We denote the binary representation of this number i by $\omega(x,y)$ and the length of this representation by ℓ, i.e. $\ell = \lceil \log m \rceil$. Also, denote

$$\mu^{(2)} = \phi_k^c(H_0)\phi_k^c(H_1)\dots\phi_k^c(H_{m-1}) \text{ and}$$

$$\mu^{(3)} = \omega(0,0)\omega(0,1)\dots\omega(0,p-1)\omega(1,0)\dots\omega(p-1,p-1).$$

The word $\mu^{(2)}$ describes all graphs H_i and the word $\mu^{(3)}$ indicates for each pair x, y the interval in the word $\mu^{(2)}$ containing the information about $G'_{x,y}$. Therefore, the words $\mu^{(2)}$ and $\mu^{(3)}$ completely describe all graphs $G_{x,y}$. In order to separate the word $\mu^{(2)}\mu^{(3)}$ into $\mu^{(2)}$ and $\mu^{(3)}$, it suffices to know the number ℓ, because $|\mu^{(2)}| = \ell p^2$ and the number p is uniquely defined by n. Since $m \le 2^{\binom{k}{2}}$, the number ℓ can be described by at most

$$\lceil \log_2 \ell \rceil = \lceil \log_2 \lceil \log_2 m \rceil \rceil \le \lceil \log_2 \binom{k}{2} \rceil \le \lceil \log_2 k^2 \rceil \le \lceil \log_2 \log_2 n \rceil$$

binary bits. Let $\mu^{(0)}$ be the binary representation of the number ℓ of length $\lceil \log \log n \rceil$, and let

$$\phi_n^*(G) = \mu^{(0)}\mu^{(1)}\mu^{(2)}\mu^{(3)}, \qquad \Phi^* = \{\phi_n^* \ : \ n = 2,3,\dots\}.$$

Theorem 2.4.1. *Φ^* is an asymptotically optimal coding for any hereditary class X with $c(X) > 1$.*

Proof. From the construction of Φ^* it is clear that any graph is uniquely defined by its code. Therefore, Φ^* is a coding for any class of graphs. Assume now that our graph G belongs to a hereditary class X with $c(X) > 1$. We denote the entropy of X by $h(X)$, i.e.

$$h(X) = \lim_{n\to\infty} \frac{\log_2 |X_n|}{\binom{n}{2}},$$

and therefore,

$$|X_n| = 2^{\frac{n^2}{2}(h(X)+\varepsilon_n)},$$

where $\varepsilon_n \to 0$ when $n \to \infty$.

Let n be the number of vertices of G. We estimate the lengths of the words in the code $\phi_n^*(G)$ as follows:

$$|\mu^{(0)}| = \lceil \log_2 \log_2 n \rceil,$$

$$|\mu^{(1)}| = k\binom{p}{2} + kp(n-kp) + \binom{n-kp}{2}.$$

Taking into account (2.2), we conclude that

$$|\mu^{(1)}| \le \frac{3kp^2}{2} + \frac{p^2}{2} \le \frac{6n^2}{\sqrt{\log_2 n}} + \frac{2n^2}{\log_2 n} = o(n^2).$$

Each graph H_i belongs to X_k and hence the number m of these graphs satisfies

$$m \le |X_k| = 2^{\frac{k^2}{2}(h(X)+\varepsilon_k)},$$

where $\varepsilon_k \to 0$ when $k \to \infty$. Therefore,

$$|\mu^{(2)}| = m\binom{k}{2} < k^2 2^{\frac{k^2}{2}(h(X)+\varepsilon_k)} \le n^{\frac{1}{2}(h(X)+\varepsilon_k)} \log_2 n,$$

$$|\mu^{(3)}| = p^2\lceil \log_2 m \rceil \le \frac{p^2k^2}{2}(h(X)+\varepsilon_k).$$

Since $h(X) \le 1$ and $k \to \infty$ when $n \to \infty$, we conclude that $|\mu^{(2)}| = o(n^2)$. Also, since $kp \le n$, we have

$$|\mu^{(3)}| \le \frac{n^2}{2}h(X) + o(n^2).$$

Combining the above arguments, we obtain

$$|\phi_n^*(G)| \le \frac{n^2}{2}h(X) + o(n^2).$$

Therefore, if $c(X) > 1$ (i.e. if $h(X) > 0$), then

$$\lim_{n\to\infty} \frac{\max\limits_{G\in X_n} |\phi_n^*(G)|}{\log_2 |X_n|} = 1,$$

and hence, Φ^* is an asymptotically optimal coding for X. □

Example 2.4.2. Let G be the graph represented in Figure 2.3. Its canonical code is

00111000 1111000 111000 10100 1001 001 10 1.

This code is obtained by listing the elements of the adjacency matrix above the main diagonal. For convenience, the elements from different rows of the matrix are separated. Let us define $k = p = 3$ (this choice of k and p does not satisfy their definition, but it is not very important. These numbers still satisfy (2.2)).

The set R_2 consists of the pairs of vertices that belong to the subgraphs of G induced by the three sets $\{1,2,3\}$, $\{4,5,6\}$ and $\{7,8,9\}$. The word $\mu^{(1)}$ consists of the elements of the canonical code corresponding to these pairs:

$$\mu^{(1)} = 001101101.$$

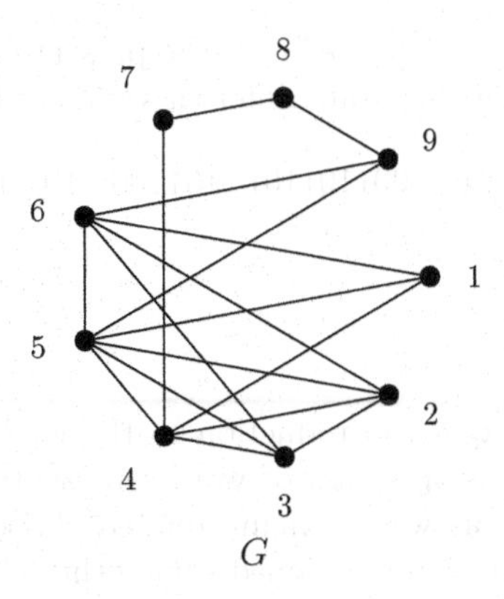

Figure 2.3: The graph G of Example 2.4.2

x, y	$Q_{x,y}$	$\phi_3^c(G'_{x,y})$	$\omega(x, y)$
0,0	1,4,7	101	0
0,1	2,5,8	100	1
0,2	3,6,9	101	0
1,0	1,5,9	101	0
1,1	2,6,7	100	1
1,2	3,4,8	100	1
2,0	1,6,8	100	1
2,1	2,4,9	100	1
2,2	3,5,7	100	1

Table 2.1: The sets $Q_{x,y}$ and canonical codes of the graphs $G'_{x,y}$

Now for each pair $(x, y) \in \{0, 1, 2\}^2$, we compute the set $Q_{x,y}$ and the canonical code of the graph $G'_{x,y}$. The results are presented in Table 2.1.

The third column of the table shows that among the graphs $G'_{x,y}$ for various values of x and y there are only two different graphs, i.e. $m = 2$ and hence $\ell = 1$. The canonical codes of these graphs are 101 and 100. These two codes give us the word $\mu^{(2)}$:

$$\mu^{(2)} = 101100.$$

We code the graph with canonical code 101 by 0 and the graph with canonical code 100 by 1. This coding gives us the word $\mu^{(3)}$:

$$\mu^{(3)} = 010011111.$$

Thus, the code of $\phi_9^*(G) = \mu^{(0)}\mu^{(1)}\mu^{(2)}\mu^{(3)}$ is represented by the following binary sequence (for convenience, we separate different words of the code):

$$\phi_9^*(G) = 01\ \ 001101101\ \ 101100\ \ 010011111.$$

2.5 Conclusion

In this chapter, we gave a quick introduction to the world of hereditary classes of graphs. This world contains the class of word-representable graphs, which is the main subject of this book, as well as many related classes, such as comparability graphs or circle graphs. We also mentioned some other classes of interest that will appear later in the book such as interval graphs, permutation graphs, line graphs etc.

The world of hereditary classes is not exhausted by the examples given in this chapter and includes, for instance, *threshold graphs, co-graphs, trapezoid graphs, chain graphs, weakly chordal graphs, convex graphs*, etc. Moreover, through the induced subgraph characterization of hereditary classes it is not difficult to see that the family of hereditary classes contains uncountably many members. Indeed, create a member of this family by forbidding a finite collection of cycles. Clearly, any two different collections of forbidden cycles lead to two different hereditary classes (see Theorem 2.1.5) and since the number of such collections is uncountable, then so is the number of hereditary classes.

In the concluding part of the chapter, we also touched on the coding of graphs, i.e. representing graphs by words in a finite alphabet. The question of optimal coding of graphs in a hereditary class X is closely related to the question of the speed of X, i.e. the number of n-vertex labelled graphs in X. For classes with high speed (i.e. non-zero entropy) an asymptotically optimal coding can be obtained using the universal algorithm described in this chapter. In the subsequent chapters, we will show that word-representable graphs constitute a hereditary class with a high speed, and therefore the algorithm given in the present chapter provides an asymptotically optimal coding for graphs in this class.

2.6 Exercises

1. Which of the following sets of graphs are hereditary?

    - complete graphs,
    - paths,

- cycles,
- connected graphs,
- graphs of diameter k,
- graphs of vertex degree at most k,
- graphs of independence number at least k.

2. Is the set of minimal forbidden induced subgraphs for the class of graphs of vertex degree at most k finite or infinite?

3. Show that the claw $K_{1,3}$ is a minimal non-line graph.

4. Find the number of vertices and edges of the line graph $L(G)$ of a graph G with the degree sequence $(d_1, d_2, \ldots, d_n)$.

5. Find the set of minimal forbidden induced subgraphs for the class of bipartite graphs every connected component of which is complete bipartite.

6. Provide a structural characterization for the class of (P_4, C_4)-free bipartite graphs.

7. Show that every $2K_2$-free bipartite graph is a permutation graph.

8. Show that the split graphs and bipartite graphs are perfect using only the definition of these classes.

9. An interval graph is *unit interval* if all intervals in the intersection model of this graph have the same length. Show that the claw $K_{1,3}$ is a minimal interval graph which is not unit interval, i.e. the class of unit interval graphs is a subclass of $K_{1,3}$-free interval graphs. What about the inverse inclusion?

2.7 Solutions to Selected Exercises

4. Each vertex in the line graph is obtained from an edge in the original graph G. In the original graph there are exactly $\frac{1}{2}\sum_{i=1}^{n} d_i$ edges. Thus the number of vertices in the line graph is $\frac{1}{2}\sum_{i=1}^{n} d_i$.

   An edge in the line graph is obtained if it connects two vertices which come from edges of G which meet at a vertex. So consider the vertex i. There are d_i edges of G meeting at i, and so $\binom{d_i}{2}$ pairs among them. Thus this vertex gives rise to $\binom{d_i}{2}$ edges in $L(G)$. Hence the line graph contains $\sum_{i=1}^{n}\binom{d_i}{2}$ edges.

5. The class of bipartite graphs every connected component of which is complete bipartite is precisely the class of P_4-free bipartite graphs. Indeed, P_4 is not complete bipartite, and hence bipartite graphs every connected component of which is complete bipartite are P_4-free. On the other hand, if a connected bipartite graph G is P_4-free, then G is complete bipartite, because any path connecting two non-adjacent vertices in different parts of the bipartition of G has at least three edges, i.e. contains an induced P_4.

Chapter 3

What Word-Representable Graphs Are and Where They Come from

Definition 3.0.1. For a finite word w, $A(w)$ denotes the set of letters occurring in w.

Example 3.0.2. For the word $w = 314836673$, $A(w) = \{1, 3, 4, 6, 7, 8\}$.

The notion of alternating letters in a word, to be defined next, is a key concept in defining word-representable graphs.

Definition 3.0.3. Suppose that w is a word and x and y are two distinct letters in w. We say that x and y *alternate* in w if after deleting in w all letters but the copies of x and y we either obtain a word $xyxy\cdots$ (of even or odd length) or a word $yxyx\cdots$ (of even or odd length). If x and y do not alternate in w, we say that these letters are *non-alternating* in w.

Note that by definition, if w has a single occurrence of x and a single occurrence of y, then x and y alternate in w.

Example 3.0.4. If $w = 31341232$ then Table 3.1 shows whether or not a pair of letters in w is alternating. For example, 1 and 3 are alternating in w because removing all other letters we obtain 31313, while 3 and 4 are non-alternating because removing all other letters from w we have 3343.

An important object in this book is defined as follows.

Definition 3.0.5. A simple graph $G = (V, E)$ is *word-representable* if there exists a word w over the alphabet $A(w) = V$ such that letters x and y alternate in w if and only if $xy \in E$, that is, x and y are connected by an edge, for each $x \neq y$. We say that w *represents* G, and w is called a *word-representant* for G.

	1	2	3	4
1	N/A	non-alternating	alternating	alternating
2	non-alternating	N/A	non-alternating	non-alternating
3	alternating	non-alternating	N/A	non-alternating
4	alternating	non-alternating	non-alternating	N/A

Table 3.1: Alternation of pairs of letters in the word $w = 31341232$

Remark 3.0.6. In this book, we typically have $V = \{1, \ldots, n\}$ for an appropriate n.

Example 3.0.7. The graphs in Figure 3.1 are word-representable. Indeed, for example, 1213423 is a word-representant for M, 1234 is a word-representant for K_4, and a word-representant for the Petersen graph is

$$1387296(10)7493541283(10)7685(10)194562. \tag{3.1}$$

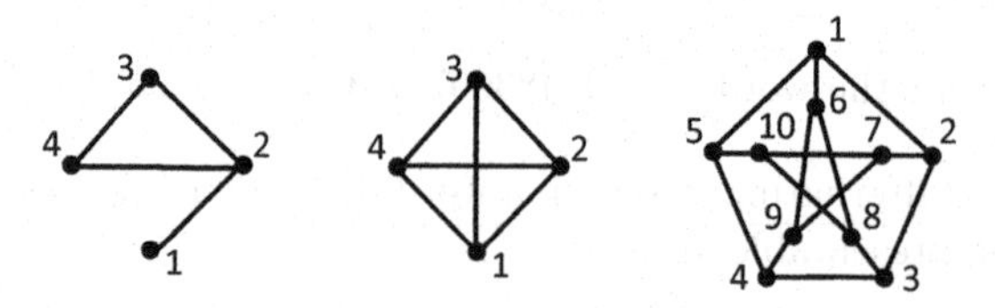

Figure 3.1: Three word-representable graphs M (left), the complete graph K_4 (middle), and the Petersen graph (right)

The following statement is easy to see by our definitions.

Proposition 3.0.8. *The class of word-representable graphs is hereditary.*

Proof. Suppose that a word w represents a graph $G = (V, E)$ and $x \in V$. Also, suppose that G' is obtained from G by removing from it the vertex x and all edges incident to x. Finally, suppose that w' is obtained from w by removing all occurrences of the letter x. Then our definitions ensure that w' represents G', so the class of word-representable graphs is hereditary. □

As we will see shortly in Proposition 3.2.13, any word-representable graph has infinitely many words representing it. For example, some of other representations of K_4 are 3241, 123412, 12341234, 432143214321 etc.

Definition 3.0.9. For a word w, we let $G(w)$ be the graph represented by w.

Example 3.0.10. $G(23142)$ is the graph K_4 in Figure 3.1.

Remark 3.0.11. Suppose that x is an isolated vertex (with no edge incident to it) in G and $G' = G - x$ is the graph obtained by removing x from G. Then, if a word w represents G', then the word xxw represents G (in fact, xx can be inserted anywhere in w). Thus, when studying word-representability of a graph, we can ignore its isolated vertices, since it is trivial to represent them.

Definition 3.0.12. The *reverse* of the word $w = w_1 \cdots w_n$ is the word $r(w) = w_n w_{n-1} \cdots w_1$.

Example 3.0.13. For the word $w = 241533$, the reverse $r(w) = 335142$.

The following proposition is straightforward from Definition 3.0.5.

Proposition 3.0.14. *If w is a word-representant for a graph G, then the reverse $r(w)$ is also (possibly the same) a word-representant for G.*

The following proposition is also straightforward from Definition 3.0.5, but it is useful not only in verifying whether a word represents a given graph or not, but also in constructing word-representants for a given graph.

Proposition 3.0.15. *([92]) Let $w = w_1 x w_2 x w_3$ be a word representing a graph G, where w_1, w_2 and w_3 are possibly empty words, and w_2 contains no x. Let X be the set of all letters that appear only once in w_2. Then possible candidates for x to be adjacent to in G are the letters in X.*

3.1 Representing Trees

To illustrate usage of Proposition 3.0.15, and to give an example of how word-representants can be found, we show how to represent an arbitrary tree. In Theorem 4.2.1 we will introduce a general algorithm to construct a word-representant based on a *semi-transitive orientation* associated with a given word-representable graph.

The basic idea to represent a tree T is to start with any single edge xy in T and to represent it by the word $w_1 = xyxy$. The next approximation involves adding another edge incident either to x or to y. Suppose, without loss of generality, that xz is another edge in T. To add this edge to our representation, we pick any copy of x in w_1 and embrace it by two copies of the letter z. Say, we have decided to pick the leftmost x, then the word $w_2 = zxzyxy$ represents two edges xy and xz (note that y and z do not alternate in w_2, and thus yz is not an edge in T, which is

Step 1	$w_1 = 1212$
Step 2	$w_2 = 123132$
Step 3	$w_3 = 12341432$
Step 4	$w_4 = 5152341432$
Step 5	$w_5 = 656152341432$
Step 6	$w_6 = 65617572341432$
Step 7 (the tree itself)	$w_7 = 6561758782341432$

Table 3.2: Word-representants corresponding to the steps in Figure 3.2

the case since T is cycle-free). Next, we will add another edge to our representation that is incident to one of x, y or z.

More generally, suppose that k edges forming a subtree T' in T have been represented by a word w_k containing exactly two copies of each letter, and an edge ce is such that $c \notin T'$ and $e \in T'$. Thus, $w_k = w_k^1 e w_k^2 e w_k^3$ for some words w_k^1, w_k^2 and w_k^3 not containing e, and the word $w_{k+1} = w_k^1 cecw_k^2 ew_k^3$ represents T' with the edge ec added to it (alternatively, we can take $w_{k+1} = w_k^1 ew_k^2 cecw_k^3$). Indeed, the alternations of all pairs of letters not involving c are the same in w_k and w_{k+1}, while, clearly, c alternates with only e in w_{k+1}.

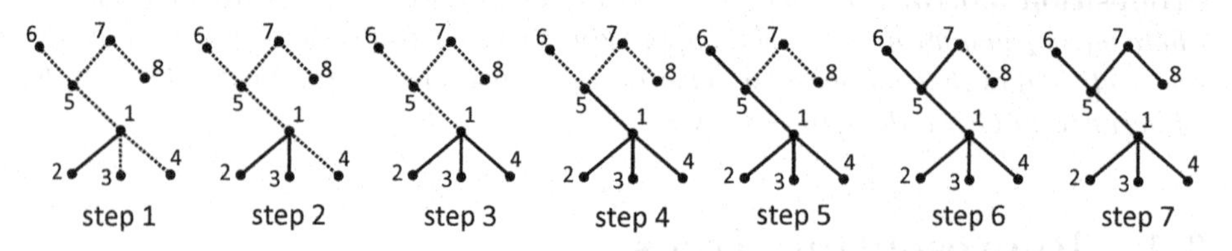

Figure 3.2: Steps in representing the tree presented in step 7

To illustrate our approach on a particular example, we show how to represent the tree to the right (presented in step 7) in Figure 3.2. The steps in Figure 3.2 can correspond to the w_is presented in Table 3.2 (notice that the choice of w_is is not unique).

3.2 k-Word-Representable Graphs

Definition 3.2.1. A word is *k-uniform* if each letter in it appears exactly k times. In particular, 1-uniform words are permutations. A word is *uniform* if it is k-uniform for some k".

Example 3.2.2. The word 243231442311 is 3-uniform.

Definition 3.2.3. A graph is k*-word-representable* if there exists a k-uniform word representing it. We say that the word k*-represents* the graph.

Example 3.2.4. In Figure 3.1, the graph M is 2-word-representable (a 2-uniform word-representant for it is 12134234), the graph K_4 is 1-word-representable (represented by, e.g. the 1-uniform word 1234), and the Petersen graph is 3-word-representable (a 3-uniform word-representant for it is given in (3.1) above). Also, as we have seen in Section 3.1, any tree is 2-word-representable.

Arguing as in the proof of Proposition 3.0.8, we see that the class of k-word-representable graphs is a hereditary class of graphs.

Definition 3.2.5. Suppose that a word $w = uv$, where u and v are two, possibly empty, words. Then the word vu is a cyclic shift of w.

Example 3.2.6. Examples of cyclic shifts of the word 24311523 are 31152324 and 52324311.

If a word represents a graph G, then its cyclic shift generally does not have to represent G. Indeed, for example, the word 121 represents the single-edge graph (the complete graph K_2), while its cyclic shift 112 represents the graph on two vertices without edges. However, in the case of k-uniform words, the property of being a word-representant is invariant with respect to cyclic shifts, as shown by the following proposition.

Proposition 3.2.7. *([92]) Let $w = uv$ be a k-uniform word representing a graph G, where u and v are two, possibly empty, words. Then the word $w' = vu$ also represents G.*

Proof. It is not difficult to see that letters x and y alternate in w if and only if they alternate in w'. □

To illustrate the usage of Proposition 3.2.7, we show how to represent an arbitrary n-cycle. Suppose that $n = 5$ and the 5-cycle is the graph to the right in Figure 3.3.

As the first step, we represent the path $1-2-3-4-5$ using the strategy to represent trees (a path is a tree) described in Section 3.1. Notice though that out of two options on a particular step, we *always* embrace the letter to the right (embracing left copies of letters would work too as long as we are consistent in doing

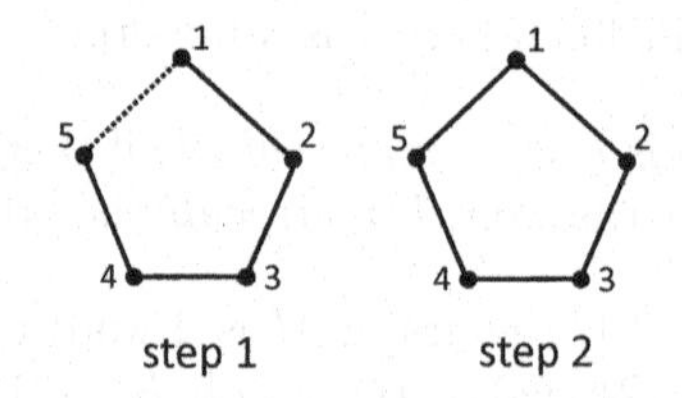

Figure 3.3: Steps in representing the 5-cycle

this). Thus, the path is represented by the word $w' = 1213243545$ which is obtained through the following steps:

$$1212 \to 121323 \to 12132434 \to 1213243545.$$

Consider the word $w'' = 5121324354$, a 1-letter cyclic shift of w', which by Proposition 3.2.7, represents the same path on five vertices. Swapping the first two letters in w'' gives the word $w = 1521324354$ representing the 5-cycle. Indeed, swapping the letters 1 and 5 does not change the alternation of any other pair of letters but creates an alternation between 1 and 5 corresponding to adding exactly one edge (between 1 and 5) to the path.

It is straightforward to generalize our construction to an arbitrary n as follows.

To represent an n-cycle:

1. Using the construction in Section 3.1, represent a path on n vertices by the word w', always embracing the second copy of the letter to be embraced;
2. In w', move the last letter to be in front of the first letter to obtain the word w'';
3. Finally, in w'' swap the first and the second letters to obtain w, the desired representation of the n-cycle.

Definition 3.2.8. A word u contains a word v as a *factor* if $u = xvy$ for possibly empty words x and y.

Example 3.2.9. The word 132243 contains the words 224 and 43 as factors, while all factors of the word 1123 are 1, 2, 3, 11, 12, 23, 112, 123 and 1123.

The following theorem provides a fact on uniform representations of cycle graphs.

Theorem 3.2.10. *([94]) The word $v = 12\cdots n$ is not a factor of any uniform representation of n-cycle C_n with vertices labelled in consecutive order, where $n \geq 4$.*

Proof. Suppose that u is a uniform word representing C_n and v is a factor of u. By Proposition 3.2.7, we can assume that v is a prefix of u, that is, $u = vw$ for some word w. Let a_i be the position of the ith occurrence of a in u (from left to right) for $a \in \{1, \ldots, n\}$. Now, for all vertices a, b connected in C_n by an edge, such that $a < b$, we have $a_i < b_i < a_{i+1}$ for each $i \geq 1$.

Vertices 1 and 3 are not adjacent in C_n, and thus they do not alternate in u. It follows that there is a $k \geq 1$ such that $1_k > 3_k$ or $3_k > 1_{k+1}$.

Suppose that $3_k < 1_k$. Since vertices 2 and 3, and 1 and 2 are adjacent in C_n, we have $2_i < 3_i$ and $1_i < 2_i$ for each i. Then we have a contradiction with $1_k < 2_k < 3_k < 1_k$. On the other hand, suppose that $1_{k+1} < 3_k$. Since all pairs of vertices $j, j+1$, as well as the pair $n, 1$, are adjacent in C_n, for each $i \geq 1$, we have that $j_i < (j+1)_i$ for each $j < n$ and $n_i < 1_{i+1}$. Thus, we obtain a contradiction with $3_k < 4_k < \cdots < n_k < 1_{k+1} < 3_k$. □

Definition 3.2.11. Suppose that $p(w)$ is the permutation obtained by removing all but the leftmost occurrence of each letter x in w. We call $p(w)$ the *initial permutation* of w.

Example 3.2.12. For $w = 21254534$, we have $p(w) = 21543$.

Proposition 3.2.13. *([92]) A k-word-representable graph G is also $(k+1)$-word-representable. In particular, each word-representable graph has infinitely many word-representants representing it since for every $\ell > k$, a k-word-representable graph is also ℓ-word-representable.*

Proof. Suppose that a k-uniform word w represents G. We claim that $p(w)w$ being a $(k+1)$-uniform word also represents G. Indeed, if letters x and y do not alternate in w then clearly they also do not alternate in $p(w)w$. Otherwise, the order of x and y in the permutation $p(w)$ is the same as that in w, and thus x and y also alternate in $p(w)w$. □

The following proposition is straightforward from our proof of Proposition 3.2.13 and from the fact that a permutation (that is, a 1-uniform word) represents a complete graph.

Proposition 3.2.14. *([94]) A k-uniform word representing a complete graph K_n is a word of the form v^k (v written k times), where v is a permutation on K_n's n vertices.*

Remark 3.2.15. Taking into account Proposition 3.2.13, it can be convenient sometimes to think of a k-uniform word-representant u of a graph to be a *cyclic word*, that is, a word written around a circle. Then, any cyclic shift of u is assumed to be equivalent to u, and u is a representative from the equivalence class with respect to the shift it defines. A benefit from dealing with cyclic words is that we do not have to treat differently the ends of a word-representant while considering a local part of it.

3.3 Graph's Representation Number and the Importance of k-Word-representability

The following theorem is of fundamental importance, showing that we can restrict ourselves to considering word-representants having the same number of copies of each letter, which is convenient, e.g. from a programming point of view.

Theorem 3.3.1. *([92]) A graph G is word-representable if and only if it is k-word-representable for some k.*

Proof. Clearly if G is k-word-representable then it is also word-representable.

Conversely, suppose a word w represents G and the maximum number of copies of a letter in w is k. To avoid triviality, we assume that w is not k-uniform.

Let the word u be obtained from w by removing all the copies of the letters appearing in w (exactly) k times. Then the word $p(u)w$, where $p(u)$ is the initial permutation of u defined in Definition 3.2.11, also represents G. Indeed, if xy is not an edge in G then x and y do not alternate in w and thus they do not alternate in $p(u)w$. On the other hand, if xy is an edge in G (x and y alternate in w), then the following (non-equivalent) cases are possible:

- $x \notin u$ and $y \notin u$: clearly x and y alternate in $p(u)w$;
- $x \in u$ and $y \in u$: by definition, the order of x and y in the permutation $p(u)$ is the same as that in w, and thus x and y alternate in $p(u)w$;
- $x \in u$ and $y \notin u$: from the definition of u, y must occur k times in w. But then x must occur $k-1$ times in w and the alternating subsequence of w induced by x and y must begin and end with y. This shows that x and y alternate in $p(u)w$.

So, the word $p(u)w$ represents G. The length of $p(u)w$ is larger than the length of w (w is not k-uniform), and the maximum number of occurrences of a letter in $p(u)w$

is still k. Thus, we can repeat the procedure above, if necessary, until we obtain a k-uniform word representing G. □

The proof of Theorem 3.3.1 suggests the following procedure for turning word-representation into uniform word-representation.

To turn a word-representant w into a k-uniform word-representant w', where k is the number of occurrences of the most frequent letter in w:

1. Consider the word u obtained from w by removing all the copies of the letters appearing in w k times;
2. Let $w := p(u)w$ where $p(u)$ is the initial permutation of u defined in Definition 3.2.11;
3. If w is k-uniform, return $w' := w$ and halt; otherwise, go to Step 2.

An immediate corollary to Theorem 3.3.1 and Proposition 3.2.13 is the following statement essentially saying that when studying word-representable graphs, one can assume that graphs are connected.

Theorem 3.3.2. *A graph G is word-representable if and only if each of G's connected components is word-representable.*

Proof. If G is word-representable then each of G's connected components is word-representable by the hereditary property of word-representable graphs.

Conversely, if each connected component of G is word-representable, then applying Theorem 3.3.1 and Proposition 3.2.13, if necessary, we can make sure that each component is represented by at least two copies of each letter. But then concatenating all words representing the components we see that the obtained word represents G (the presence of at least two copies of each letter guarantees that no edge is created between two connected components). □

Definition 3.3.3. The minimal k such that a graph is k-word-representable is called the *graph's representation number*, and it is denoted by $\mathcal{R}(G)$. Graph's representation number is also known as *graph's alternation number* (see, e.g. [74]). Also, $\mathcal{R}_k := \{G \mid \mathcal{R}(G) = k\}$.

Example 3.3.4. Referring to Example 3.2.4, $\mathcal{R}(M) = 2$ because 1-uniform words can only represent complete graphs. Also, clearly $\mathcal{R}(K_4) = 1$. Finally, for the

Petersen graph, the graph's representation number is 3 as will be shown in Theorem 5.2.1.

Remark 3.3.5. It is not difficult to see from the definitions that if $\mathcal{R}(G) = k$ and G' is an induced subgraph of G then $\mathcal{R}(G') \leq k$. Indeed, if representing G' would require more than k copies of each letter, then representing G would require more than k copies of each letter.

It turns out that $\mathcal{R}_1$ is exactly the class of complete graphs, while $\mathcal{R}_2$ is exactly the class of *circle graphs* (see Section 5.1). No characterization is known for $\mathcal{R}_k$ with $k \geq 3$. However, there are several interesting results on 3-word-representable graphs that are presented in Section 5.2. A particular property of graphs in $\mathcal{R}_k$ is recorded in the following proposition.

Proposition 3.3.6. *([89]) Let $G \in \mathcal{R}_k$, where $k \geq 2$, and $x \in V(G)$. Also, let G' be the graph obtained from G by adding an edge xy, where $y \notin V(G)$. Then $G' \in \mathcal{R}_k$.*

Proof. Suppose that G is k-represented by a word $w_0xw_1xw_2 \cdots xw_{k-1}xw_k$, where for $0 \leq i \leq k$, w_i is a word not containing x. Then it is not difficult to check that the word

$$w_0yxyw_1xw_2yxw_3yxw_4 \cdots yxw_{k-1}yxw_k$$

k-represents G' (in particular, the vertex x is the only neighbour of y). Finally, if G' could be $(k-1)$-represented by some word, we would remove from that word the letter y to obtain a $(k-1)$-representation of G, which is impossible. So, $G' \in \mathcal{R}_k$. □

In Theorem 4.2.1 we will see that the maximum graph's representation number of word-representable graphs is linear in the number of vertices. More precisely, this number cannot exceed n for a graph on n vertices, while we have an example of a graph G on n vertices with $\mathcal{R}(G) = \lfloor \frac{n}{2} \rfloor$ (see Subsection 4.2.1).

3.4 Permutationally Representable Graphs

Definition 3.4.1. A graph G with the vertex set $V = \{1, \ldots, n\}$ is *permutationally representable* if it can be represented by a word of the form $p_1 \cdots p_k$, where p_i is a permutation of V for $1 \leq i \leq k$. If G can be represented permutationally involving k permutations, we say that G is *permutationally k-representable.* In explicit representations by permutations, we arrange space between consecutive permutations for visual convenience, while one should remember that we still deal with a single word.

Example 3.4.2. A complete graph K_n is permutationally representable for any n. Indeed, take any permutation of $\{1, \ldots, n\}$ and repeat it as many times (maybe none) as desired. For another example, the path $1 - 2 - 3$ is also permutationally representable, and one such representation is 213 231. More such representations can be obtained by adjoining any number of permutations 213 and 231 to it.

The following theorem shows that comparability graphs, that is, transitively orientable graphs, are permutationally representable. It is a straightforward usage of the fact that any poset is an intersection of a number of linear (total) orders, and a linear order can be recorded as a permutation. This theorem will be extended to the case of any word-representable graphs and semi-transitive orientations in Chapter 4 (see Theorem 4.1.8).

Theorem 3.4.3. *([95]) A graph is permutationally representable if and only if it is a comparability graph.*

Proof. It follows from Remark 3.6.8, Proposition 3.6.9, and the definitions. □

As an immediate corollary to Theorem 3.4.3, we have the following statement.

Theorem 3.4.4. *Bipartite graphs are permutationally representable.*

Proof. In a given bipartite graph orient all edges from one part towards the other one to obtain a transitive orientation of the graph. Thus, any bipartite graph is a comparability graph (a well known fact), and by Theorem 3.4.3 it is permutationally representable. □

Remark 3.4.5. Finding the minimum number of permutations to be concatenated in order to represent a bipartite graph is not an easy problem. However, in some cases we can easily find this number. For example, a complete bipartite graph $K_{n,m}$ requires just two permutations. Indeed, suppose that the two parts of $K_{n,m}$ are formed by the sets $\{x_1, \ldots, x_n\}$ and $\{y_1, \ldots, y_m\}$. Then the word $x_1 \ldots x_n y_1 \ldots y_m x_n x_{n-1} \ldots x_1 y_m y_{m-1} \ldots y_1$ obtained by concatenation of two permutations (permutationally) represents $K_{n,m}$.

Remark 3.4.6. It is straightforward to see that each permutationally representable graph is k-word-representable for some k, while the converse to this statement is not true. Indeed, for example an n-cycle for odd $n \geq 5$ is not a comparability graph (and thus is not permutationally representable) while it is 2-word-representable (see Section 3.2).

Theorem 3.4.7. *([92]) Let n be the number of vertices in a graph G and $x \in V(G)$ be a vertex of degree $n-1$ (called a* dominant *or* all-adjacent *vertex). Let $H = G\backslash\{x\}$ be the graph obtained from G by removing x and all edges incident to it. Then G is word-representable if and only if H is permutationally representable.*

Proof. If H can be represented by a word $w = p_1 \cdots p_k$, where all p_i are permutations, then the word $p_1 x p_2 x \cdots p_k x$ clearly represents G (x alternates with each letter in the word, while alternation of other letters is not changed).

If G is word-representable, then by Theorem 3.3.1 and Proposition 3.2.7 it can be represented by a k-uniform word $w = xw_1xw_2 \cdots w_{k-1}xw_k$ for some words w_i not containing the letter x. Then by Proposition 3.0.15, each w_i for $i = 1, \ldots, k-1$ must be a permutation. Since w is k-uniform, w_k is also a permutation. Clearly, the word $w_1 \cdots w_k$ represents H (it has the same alternating properties of pairs of letters as those in w when x is ignored), and thus H is permutationally representable. □

Theorems 3.4.7 and 3.4.3 give an idea of the structure of word-representable graphs, which is manifested in the following theorem.

Theorem 3.4.8. *([92]) In a word-representable graph, the neighbourhood of each vertex is a comparability graph.*

Proof. It is straightforward from Theorems 3.4.7 and 3.4.3. □

A natural question to ask is as follows: if we have a graph in which the neighbourhood of each vertex is a comparability graph, then is this graph word-representable? In other words, does the converse of Theorem 3.4.8 hold? The answer to this question is no. A counterexample to the converse of Theorem 3.4.8 is given in [73] and it is so-called *co-(T_2) graph*, which is presented in Figure 3.4.

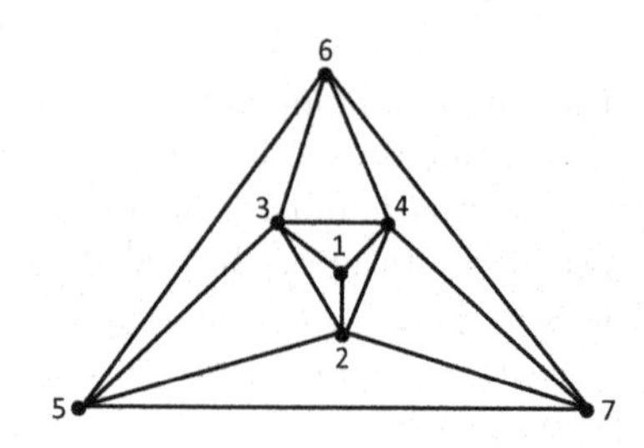

Figure 3.4: The graph co-(T_2) is a counterexample to the converse of Theorem 3.4.8

Indeed, it is easy to check that the induced neighbourhood of any vertex of co-(T_2) is a comparability graph — see Figure 3.5 for examples of transitive orientations of the neighbourhood of vertex 1, that of vertex 2 (isomorphic to the neighbourhoods of vertices 3 and 4), and the neighbourhood of vertex 5 (isomorphic to the neighbourhoods of vertices 6 and 7). However, the following theorem shows that co-(T_2) is not word-representable.

Theorem 3.4.9. *([73]) The graph co-(T_2) is not word-representable.*

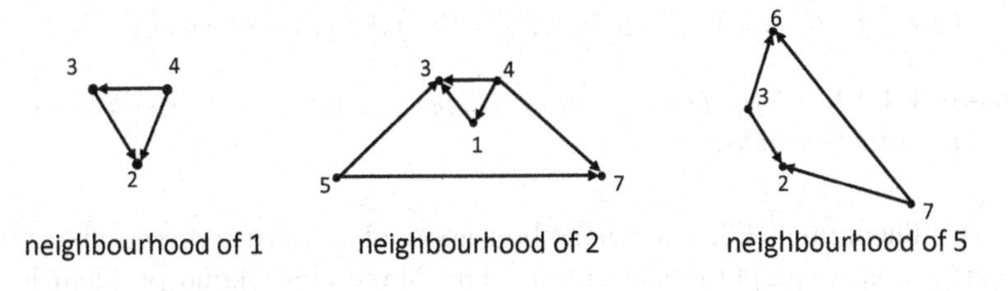

Figure 3.5: Transitive orientations of neighbourhoods of vertices in co-(T_2)

Proof. Assume that co-(T_2) is k-word-representable for some k and w is a word representing it. The vertices 1, 2, 3, and 4 form a clique in co-(T_2), and thus they must appear in w in the same order when reading w from left to right.

By Proposition 3.2.7, we can assume that 1 is the leftmost letter in w out of the letters in $\{1, 2, 3, 4\}$. Moreover, by symmetry applied to the letters 2, 3 and 4, we can assume that the order of the four letters in w is 1234. Now, let $I_1, \ldots, I_k$ be the set of all factors of w, called *intervals*, such that I_i is the factor of w between, and including, the ith occurrence of 2 and the ith occurrence of 4. Two cases are possible.

- There is an interval I_j such that 7 belongs to it. Then, since 2, 4, and 7 form a clique, 7 must be inside each of the intervals $I_1, \ldots, I_k$. But then, 7 is adjacent to 1, a contradiction.

- 7 does not belong to any of the intervals $I_1, \ldots, I_k$. Again, since 7 is adjacent to 2 and 4, each pair of consecutive intervals I_j, I_{j+1} must be separated by a single 7. But then 7 is adjacent to 3, a contradiction.

Thus, co-(T_2) is not word-representable. □

Remark 3.4.10. Another counterexample to the converse of Theorem 3.4.8 is obtained from co-(T_2) by removing the edges 23, 24 and 57 (see Figure 4.4). Indeed, the neighbourhood of each vertex of that graph has at most four vertices and thus is a comparability graph, while the proof of Theorem 4.5.1 shows that this graph is non-word-representable.

Definition 3.4.11. A *clique* in an undirected graph is a subset of pairwise adjacent vertices. A *maximum clique* is a clique of the maximum size.

Example 3.4.12. Examples of cliques in Figure 3.4 are sets formed by vertices $3, 5, 6$ or $1, 2, 3, 4$. The second of these examples is a maximum clique.

Definition 3.4.13. Given a graph G, the *Maximum Clique problem* is to find a maximum clique in G.

A direct corollary to Theorem 3.4.8 is the following statement.

Theorem 3.4.14. *([74]) The Maximum Clique problem is polynomially solvable on word-representable graphs.*

Proof. By Theorem 3.4.8, each neighbourhood of a word-representable graph G on n vertices is a comparability graph. The Maximum Clique problem is known to be solvable on comparability graphs in polynomial time [69] and thus it must be polynomially solvable on G, since any maximum clique belongs to the neighbourhood of a vertex including the vertex itself. □

3.5 Examples of Non-word-representable Graphs

In this section, we discuss known methods to construct non-word-representable graphs. In what follows, we need the following definition.

Definition 3.5.1. The *wheel graph* W_n is the graph on $n+1$ vertices obtained from the cycle graph C_n by adding an all-adjacent vertex.

Example 3.5.2. The graph W_5 is presented to the left in Figure 3.6.

3.5.1 Non-word-representable Graphs with an All-Adjacent Vertex

Theorems 3.4.3 and 3.4.7 give us a method appearing in [92] to construct non-word-representable graphs. We can take a non-comparability graph (that is, a graph having no transitive orientation; the smallest one is the cycle C_5 as is shown below), and add an all-adjacent vertex to it. Several examples of small non-word-representable graphs obtained using this method can be found in Figure 3.6. In particular, the smallest non-word-representable graph is the wheel W_5, which is the leftmost graph in Figure 3.6.

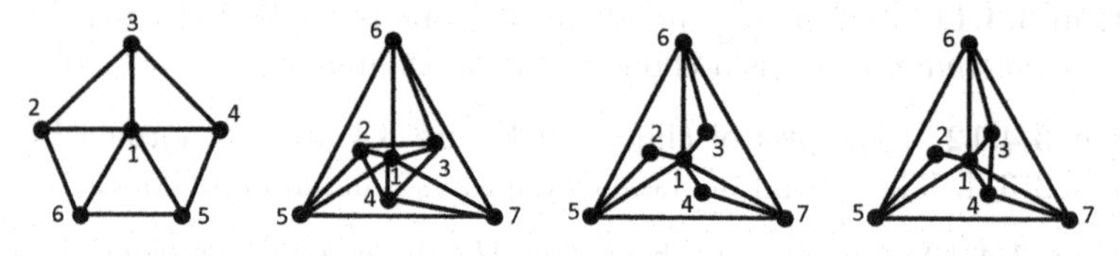

Figure 3.6: Examples of smallest non-word-representable graphs

To show that the examples in Figure 3.6 are correct, we need to show that the neighbourhoods of the all-adjacent vertices in the graphs presented in Figure 3.7 are non-comparability graphs. To this end, we show that the graphs in Figure 3.7 do not accept transitive orientations, where by $i \to j$ we denote orientation of an edge ij from i to j:

- For the 5-cycle C_5, without loss of generality, let the orientation of the edge 56 be $6 \to 5$. To be a transitive orientation, the edge 45 must be oriented as $4 \to 5$ because there is no edge between the vertices 6 and 4. Because there is no edge between 3 and 5, to have a transitive orientation, we must orient 34 as $4 \to 3$, and because there is no edge between 2 and 4, to have a transitive orientation, we must orient 23 as $2 \to 3$. However, there is no way to orient the edge 26 so that the obtained orientation is transitive, and thus C_5 is not a comparability graph.
- Using arguments similar to those used for C_5, we see that if in the graph A, without loss of generality, we have $5 \to 7$, then we also have $4 \to 7$ leading to $4 \to 3$ and $4 \to 2$. But then we must have $5 \to 2$ and $6 \to 3$ leaving us no chance to orient the edge 23 so that the obtained orientation is transitive. Thus, A in Figure 3.7 is not a comparability graph.
- For the graph B in Figure 3.7, if, without loss of generality, we have $5 \to 7$, then we also have $5 \to 2$ and $4 \to 7$. Further, $5 \to 2$ implies $5 \to 6$, and $4 \to 7$ implies $6 \to 7$. But now, there is no way to orient properly the edge 36, so that B is not a comparability graph.
- Exactly the same arguments as those for the graph B show that the graph C in Figure 3.7 is not a comparability graph.

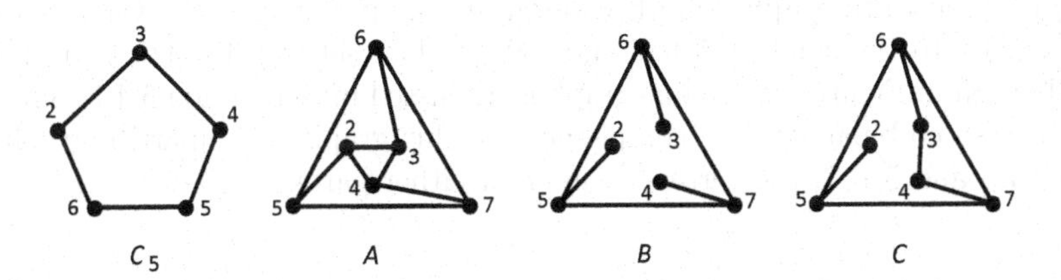

Figure 3.7: Neighbourhoods of all-adjacent vertices in the respective graphs in Figure 3.6

Conducting arguments similar to those for the cycle C_5 and the wheel W_5 above, one can see that the following proposition holds.

Proposition 3.5.3. *([92]) For any $n \geq 2$, the wheel W_{2n+1} is non-word-representable.*

Based on Figure 3.6, we can conclude that several classes of graphs are not necessarily word-representable, which, in particular, explains some of relations in Figure 1.1.

Proposition 3.5.4. *The following classes of graphs contain both word-representable and non-word-representable graphs:*

- *4-colourable graphs;*
- *4-colourable and K_4-free graphs;*
- *perfect graphs;*
- *chordal graphs;*
- *split graphs;*
- *line graphs;*
- *co-bipartite graphs.*

Proof. Our proof is straightforward from the following facts. All four graphs in Figure 3.6 are 4-colourable, while the first graph, the wheel W_5, is also K_4-free. All but the first graph in Figure 3.6 are perfect (they contain the 4-clique on vertices $1, 5, 6, 7$, while the second graph additionally contains the 4-clique on vertices $1, 2, 3, 4$). The third graph is both chordal and a split graph (the vertices can be partitioned into the clique $1, 5, 6, 7$ and independent set $2, 3, 4$). The fact that the second graph in Figure 3.6 is a line graph is discussed in Subsection 5.4.7. The same graph is also co-bipartite (the complement of this graph is a bipartite graph with parts $2, 3, 4$ and $5, 6, 7$; the vertex 1 can enter either part). □

Let us emphasize that the graphs shown in Figure 3.6 are *minimal* non-word-representable. Since the class of word-representable graphs is hereditary, any graph containing a non-word-representable graph as an induced subgraph is also non-word-representable. For instance, the graph of Figure 3.8 is not word-representable, because it contains the wheel W_5 as an induced subgraph.

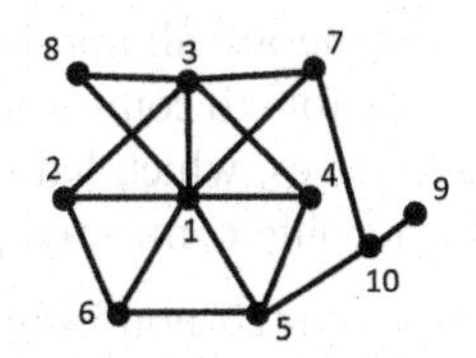

Figure 3.8: An example of a non-word-representable graph

3.5.2 Non-word-representable Line Graphs

Non-word-representable line graphs were mentioned in the proof of Proposition 3.5.4. In Section 5.5, we will extend our knowledge about non-word-representable line graphs. Namely, we will show that for any $n \geq 4$, the line graph $L(W_n)$ is not word-representable (see Theorem 5.5.4), and for $n \geq 5$, the line graph $L(K_n)$ is not word-representable (see Theorem 5.5.5). Moreover, Theorem 5.5.6 will show that if a connected graph G is not a path, a cycle or the claw graph $K_{1,3}$, then the line graph $L^n(G)$ is not word-representable for $n \geq 4$.

3.5.3 More on Non-word-representable Graphs

Non-word-representable graphs constructed using the method in Subsection 3.5.1 have the property that all of them contain a triangle. Indeed, a triangle is necessarily created when adding an all-adjacent vertex to a graph containing at least one edge. A natural question is whether there are any triangle-free non-word-representable graphs. Theorem 4.4.1 in Section 4.4 gives a positive answer to this question and shows how to construct an infinite family of such graphs.

Another observation is that a non-comparability graph must have at least five vertices, and thus any non-word-representable graph constructed using the method in Subsection 3.5.1 necessarily has a vertex of degree at least 5. It was an open question for several years whether there exist non-word-representable graphs of maximum degree 4. Theorem 4.5.1 in Section 4.5 answers this question in the affirmative.

According to experiments run by Herman Z.Q. Chen [34], there are 1, 25 and 929 non-isomorphic non-word-representable connected graphs on six, seven and eight vertices, respectively. These numbers were confirmed and extended to 68,545 for nine vertices, and 4,880,093 for 10 vertices, using a constraint programming (CP)-based method by Özgür Akgün, Ian Gent and Christopher Jefferson. Using the CP solver *Minion* [65] and the CP modelling tool *Savile Row* [113], they constructed a CP model to count the number of word-representable graphs in a given set of candidate graphs. Using their model in combination with Brendan McKay and

Adolfo Piperno's geng tool [109] to generate all non-isomorphic connected graphs for a given number of vertices, they were able to count the number of word-representable connected graphs with six to ten vertices, which led to the results mentioned above, since the total number of non-isomorphic connected graphs is known.

Figure 3.9 created by Chen shows 25 non-isomorphic non-word-representable graphs on seven vertices. Note that the only non-word-representable graph on six vertices is the wheel W_5. Further note that the case of seven vertices gives just 10 minimal non-isomorphic non-word-representable graphs, since 15 of the graphs in Figure 3.9 contain W_5 as an induced subgraphs (these graphs are the first 11 graphs, plus the 15th, 16th, 18th and 19th graphs).

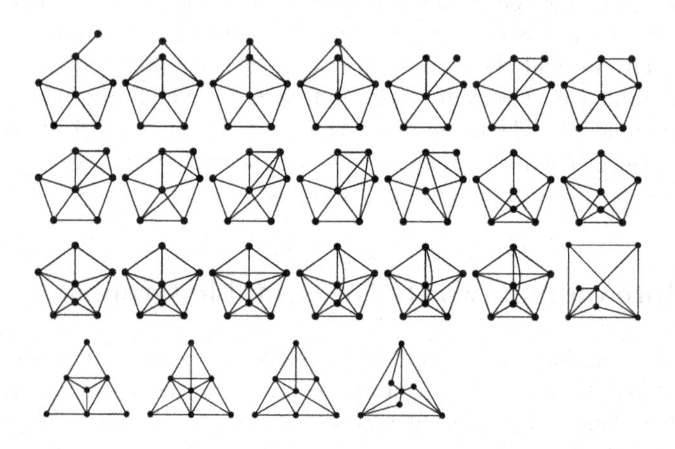

Figure 3.9: 25 non-isomorphic non-word-representable graphs on seven vertices

3.6 Some History: Study of the Perkins Semigroup

Kitaev introduced word-representable graphs in 2004. This was influenced by *alternation word digraphs* used by Kitaev and Seif [95] as a tool to study the celebrated *Perkins semigroup*, $\mathbf{B}_2^1$, which has played a central role in semigroup theory since 1960, particularly as a source of examples and counterexamples. However, the first systematic study of word-representable graphs was not undertaken until the appearance in 2008 of the paper [92] by Kitaev and Pyatkin, which started the development of the theory of word-representable graphs.

A major contribution to the theory was made in 2010 by Halldórsson, Kitaev and Pyatkin [73, 74, 75] who not only solved a number of previously stated open problems, but also provided a characterization of word-representable graphs in terms of so-called *semi-transitive orientations* (see Chapter 4). In particular, among several other interesting results based on the characterization, it was shown in [74, 75] that 3-colourable graphs are word-representable (see Section 4.3). Also, in 2010, Konovalov and Linton used the constraint solver Minion (see http://minion.sourceforge.net) to find two 3-uniform words representing the Petersen graph, thus disproving a conjecture in [92] on non-word-representability of this graph (at the time of the findings of Konovalov and Linton, it was not known that all 3-colourable graphs are word-representable).

Further developments in the study of word-representable graphs were made in 2011 by Kitaev, Salimov, Severs and Úlfarsson [93, 94], who considered the graphs in relation to the taking the line graph operation, and in 2013 by Collins, Kitaev and Lozin [40], who provided an asymptotic enumeration of word-representable graphs based on a general result on hereditary classes of graphs. Recent results due to Akrobotu, Kitaev and Masárová [3], made available in 2014, concern word-representability of polyomino triangulations. Some of these results were generalized by Glen and Kitaev [68] in 2015.

While Chapters 3–7 are dedicated to a detailed exposition of the state of the art in the field of word-representable graphs (including directions for further research), we do not provide much detail on alternation word digraphs and the study of the Perkins semigroup, since this is not directly related to our main interests in the book. Nevertheless, in Subsection 3.6.1 we introduce alternation word digraphs and briefly discuss their significance in the study of two problems (the *word problem* and the *estimation of free spectrum problem*) on the Perkins semigroup in Subsections 3.6.2 and 3.6.3, respectively. In presenting the material in the rest of the section, we follow [95].

3.6.1 Alternation Word Digraphs

The following definition generalizes Definition 3.0.3.

Definition 3.6.1. Let w be a word over an alphabet A, and X and Y be disjoint non-empty subsets of A. Then X and Y are *alternating* in w if by deleting all letters in w but those in $X \cup Y$ we obtain a word in which the letters in odd positions are from X and the letters in even positions are from Y, or vice versa.

Example 3.6.2. The sets $\{1, 5\}$ and $\{3, 4\}$ alternate in the word 31235412.

Definition 3.6.3. For a given word w over A, its *alternation word digraph* $Alt(w)$ has the vertex set consisting of the non-empty proper subsets of A. For two disjoint non-empty proper subsets $X, Y \subset A$, we let $X \to Y$ be an edge of $Alt(w)$ if and only if

- X and Y are alternating in w, and
- the leftmost letter in w from $X \cup Y$ belongs to X.

Example 3.6.4. In Figure 3.10, we provide the alternation word digraph $Alt(w)$ corresponding to the word $w = 12135134$, where we skip writing parentheses and commas in all the sets involved. In particular, there is an edge from the set $\{1,4\}$ to the set $\{2,3\}$ because by removing from w all letters but those in $\{1,2,3,4\}$ we obtain the word 1213134 having the right alternating properties, and the leftmost letter in the word is $1 \in \{1,4\}$.

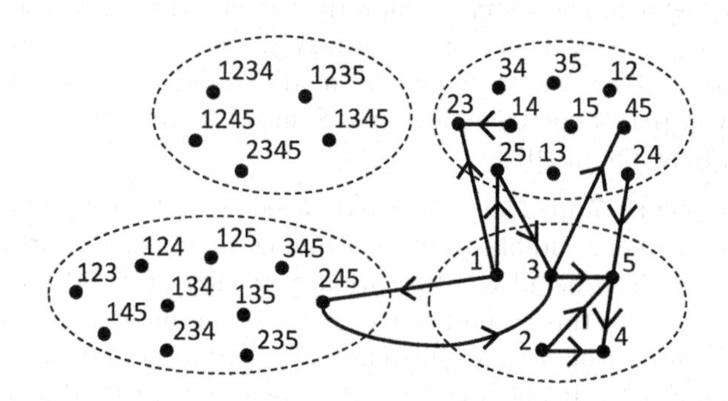

Figure 3.10: The alternation word digraph corresponding to the word 12135134

It is easy to show [95] from the definitions that an alternation word digraph is a *directed acyclic graph* (*DAG*).

Definition 3.6.5. Let $Alt_1(w)$ be the induced subgraph of $Alt(w)$ on the level of singletons (sets containing a single letter).

Example 3.6.6. $Alt_1(w)$ for the graph in Figure 3.10 is formed by the vertices $1, \dots, 5$ and the four edges between these vertices.

Theorem 3.6.7. *([95]) If $\mathbf{P} = \langle X; <_{\mathbf{P}} \rangle$ is a finite labelled poset, then there exists a finite word w whose set of letters is X such that $Alt_1(w)$, after removing the orientations of the edges, is the comparability graph corresponding to $\mathbf{P}$.*

Proof. Let $X = \{1, \dots, n\}$ and $\mathbf{P}$ have a realizer $\mathcal{R} = \{L_1, \dots, L_k\}$. For $i = 1, \dots, k$, we have L_i is a total ordering $\sigma_1^i > \cdots > \sigma_n^i$, where $\sigma^i = \sigma_1^i \cdots \sigma_n^i$ is a permutation of $\{1, \ldots, n\}$. Let $w_{\mathcal{R}} = \sigma_1^1 \cdots \sigma_n^1 \cdots \sigma_1^k \cdots \sigma_n^k$ be a word of length kn in the alphabet $\{1, \ldots, n\}$. It is not difficult to see that for $i, j \in \{1, \ldots, n\}$, $i \neq j$, we have $i <_{\mathbf{P}} j$ if and only if i and j alternate in $w_{\mathcal{R}}$, and the leftmost occurrence of i precedes the leftmost occurrence of j in $w_{\mathcal{R}}$. That is, $i <_{\mathbf{P}} j$ if and only if $i \to j \in Alt(w_{\mathcal{R}})$, so that $Alt(w_{\mathcal{R}})$, after removing the orientations of the edges, is the comparability graph corresponding to the poset $\mathbf{P}$. □

Remark 3.6.8. It is not difficult to see from Theorem 3.6.7 that any comparability graph is the orientation-free $Alt_1(w)$ for some word w written as in Definition 3.4.1 using a number of permutations next to each other. Conversely, any such word w defines a comparability graph via the orientation-free $Alt_1(w)$.

The following proposition is easy to see from the definitions.

Proposition 3.6.9. *Removing the orientations of edges in $Alt_1(w)$, we obtain a word-representable graph. Conversely, for any word-representable graph G, there exists a word w such that G is obtained from $Alt_1(w)$ by removing the orientations of edges.*

Proof. The first statement is straightforward from the definitions of $Alt_1(w)$ and word-representable graphs after not paying attention to which of the alternating letters appears first in w. As for the second statement, take any word w representing G and remove the orientations of edges in $Alt_1(w)$ to obtain G. □

3.6.2 Word Problem for the Perkins Semigroup

Recall that a *semigroup* $\mathbf{S} = \langle S; * \rangle$ is a set S equipped with an associative binary operation $*$. For $a, b \in S$, we write "ab" rather than "$a*b$". A *monoid* is a semigroup with an identity element $\mathbf{1}$, satisfying $\mathbf{1}a = a = a\mathbf{1}$.

Definition 3.6.10. The six-element monoid $\mathbf{B_2^1} = \langle B_2^1; \cdot \rangle$, the Perkins semigroup, has the following elements with the usual matrix multiplication operation:

$$\mathbf{0} = \begin{pmatrix} 0 & 0 \\ 0 & 0 \end{pmatrix}, \ \mathbf{1} = \begin{pmatrix} 1 & 0 \\ 0 & 1 \end{pmatrix}, \ a = \begin{pmatrix} 0 & 0 \\ 1 & 0 \end{pmatrix}, \ a' = \begin{pmatrix} 0 & 1 \\ 0 & 0 \end{pmatrix},$$

$$aa' = \begin{pmatrix} 0 & 0 \\ 0 & 1 \end{pmatrix}, \ a'a = \begin{pmatrix} 1 & 0 \\ 0 & 0 \end{pmatrix}.$$

To make a connection to algebra, in what follows we refer to the elements of the alphabet $\{x_1, x_2, \ldots\}$ as *variables* rather than letters, and think of a word as a non-commutative product of its variables.

Definition 3.6.11. Let $Var(w) \subset \{x_1, x_2, \ldots\}$ denote the variables occurring in a finite word w. Also, let $V_1(w)$ be the subset of $Var(w)$ consisting of variables that occur exactly once in w.

Example 3.6.12. For the word $w = x_2x_4x_6x_5x_8x_5$, $Var(w) = \{x_2, x_4, x_5, x_6, x_8\}$ and $V_1(w) = \{x_2, x_4, x_6, x_8\}$.

Remark 3.6.13. From Definitions 3.0.1 and 3.6.11, $A(w) = Var(w)$.

Definition 3.6.14. For a word w, let $Var(w) \subseteq \{x_1, \ldots, x_n\}$. Then $\overline{w}$ is the word over the alphabet $\{x_1, \ldots, x_n\} \cup \{y_1, \ldots, y_n\}$ formed from w by replacing each instance $x_i \in Var(w)$ with x_iy_i.

Example 3.6.15. For the word $w = x_2x_1x_1x_4x_6$, $\overline{w} = x_2y_2x_1y_1x_1y_1x_4y_4x_6y_6$.

Definition 3.6.16. Let $w = x_{i_1} \cdots x_{i_j}$ be a word for some $j \geq 1$, and let X be a set of variables containing $Var(w)$, that is, $Var(w) \subseteq X$. An *evaluation* $e : X \to B_2^1$ is an assignment $x_i \to s_i \in B_2^1$ for all $x_i \in X$. Let $e(w) = e(x_{i_1}) \cdots e(x_{i_j})$ be the *evaluation of w under e*.

Example 3.6.17. With $w = x_2x_1x_2$ and the evaluation $e : Var(w) = \{x_1, x_2\} \to B_2^1$ given by $e(x_1) = a$ and $e(x_2) = a'$, we have $e(w) = a'aa' = a'$.

Similarly, evaluations can be defined for any semigroup **S**.

Definition 3.6.18. If u and v are words such that for all evaluations $e : Var(u) \cup Var(v) \to S$ we have $e(u) = e(v)$, then u and v are said to be **S**-*equivalent*, which is denoted by $u \approx_{\mathbf{S}} v$. The expression $u \approx_{\mathbf{S}} v$ is said to be an *identity* of **S**.

Example 3.6.19. A semigroup **S** is commutative if and only if $x_1x_2 \approx_{\mathbf{S}} x_2x_1$.

It is not difficult to verify that $x_1^2x_2^2 \approx_{\mathbf{B_2^1}} x_2^2x_1^2$ and that $x_1^3 \approx_{\mathbf{B_2^1}} x_1^2$, two $\mathbf{B_2^1}$ identities that played a key role in [95]. The importance of the monoid $\mathbf{B_2^1}$ was established when, in 1969, Perkins [114] proved that there exists no finite set Γ of $\mathbf{B_2^1}$-identities such that all identities of $\mathbf{B_2^1}$ can be derived from Γ. That is, $\mathbf{B_2^1}$ does not have a *finite basis for its identities.*

The word problem for a semigroup S: For given finite semigroup **S** and words u and v, is $u \approx_{\mathbf{S}} v$?

The word problem for a finite semigroup **S** is decidable. The main theorem in [95], a solution to the word problem for $\mathbf{B_2^1}$ in terms of alternation word digraphs, is stated next, where recall that $r(w)$ is the reverse of the word w defined in Definition 3.0.12.

Theorem 3.6.20. *([95]) Let u and v be words. Then $u \approx_{\mathbf{B}_2^1} v$ over $\mathbf{B}_2^1$ if and only if*

1. *$V_1(u) = V_1(v)$ (see Definition 3.6.11),*
2. *$Alt(\overline{u}) = Alt(\overline{v})$, (see Definitions 3.6.3 and 3.6.14), and*
3. *$Alt(\overline{r(u)}) = Alt(\overline{r(v)})$ (see Definitions 3.6.3, 3.0.12 and 3.6.14).*

3.6.3 Free Spectra of the Perkins Semigroup

Let $\mathbf{M} = \langle M; * \rangle$ be a finite monoid, and let w be a finite word such that $Var(w) \subseteq \{x_1, \ldots, x_n\}$. Then w determines an n-ary function $w^{\mathbf{M}} : M^n \to M$ where for $(a_1, \ldots, a_n) \in M^n$, we let $w^{\mathbf{M}}(a_1, \ldots, a_n) = e(w)$, where $e : \{x_1, \ldots, x_n\} \to M$ is the evaluation (see Definition 3.6.16) such that $e(x_i) = a_i$, for $i = 1, \ldots, n$. If $i \leq n$ and $x_i \notin Var(w)$, then x_i is an *inessential variable* in the function $w^{\mathbf{M}}$. Observe that $u \approx_{\mathbf{M}} v$ if and only if $u^{\mathbf{M}} = v^{\mathbf{M}}$.

Definition 3.6.21. Let $F_n(\mathbf{M}) = \{w^{\mathbf{M}} : Var(w) \subseteq \{x_1, \ldots, x_n\}\}$, $f_n^{\mathbf{M}}$ be the cardinality of $F_n(\mathbf{M})$, and let

$$(f_n^{\mathbf{M}})_{n \geq 1} = (f_1^{\mathbf{M}}, f_2^{\mathbf{M}}, \ldots)$$

denote the *free spectrum* of $\mathbf{M}$.

Example 3.6.22. With $\mathbf{U_2}$, the two-element semilattice consisting of 0 and 1 with ordinary multiplication, we have $f_n^{\mathbf{U_2}} = 2^n - 1$. Indeed, non-empty subsets of $\{x_1, \ldots, x_n\}$ listed as words (order of variables in a word is irrelevant) will give $2^n - 1$ distinct functions, while adding copies of an already used variable will not create a new function, since $x_i^2 = x_i$ for all i.

Note that $f_n^{\mathbf{M}} \leq |M|^{|M|^n}$. Free spectra of a finite monoid $\mathbf{M}$ are typically analyzed up to $O(\log f_n^{\mathbf{M}})$. A free spectrum is said to be *log-exponential* (or, *doubly exponential*) if there exists a finite real number c such that for all n high enough $f_n^{\mathbf{M}} \geq 2^{2^{cn}}$; otherwise, $\mathbf{M}$ is said to be *sub-log-exponential.*

The free spectrum problem for a finite monoid M:
Is the free spectrum of $\mathbf{M}$ sub-log-exponential? If so, determine $O(\log f_n^{\mathbf{M}})$.

The following theorem, proved in [95], gives upper and lower bounds for the elements of the free spectrum of $\mathbf{B}_2^1$ in terms of a_n, the number of distinct alternation word digraphs on words whose alphabet is contained in $\{x_1, \ldots, x_n\}$.

Theorem 3.6.23. *([95]) For all $n \geq 1$, we have*

$$p_n \leq a_n \leq f_n^{\mathbf{B}_2^1} \leq 2^n a_{2n}^2,$$

where p_n is the number of distinct labelled posets on $\{1, \ldots, n\}$.

A result of Seif [125] states that log $a_n \in O(n^3)$. But then log $2^n a_{2n}^2 \in O(n^3)$ and from Theorem 3.6.23, it follows that log $f_n^{\mathbf{B}_2^1} \in O(n^3)$ as well.

3.7 Conclusion

In this chapter we introduced the key notion of this book, namely that of a word-representable graph, influenced by the study of the celebrated Perkins semigroup. Not all graphs are word-representable, with the wheel W_5 being the smallest non-word-representable graph, but the class of word-representable graphs is hereditary. A useful observation is that any word-representable graph can be represented by a uniform word, that is, by a word containing the same number of occurrences of each letter. This leads to the notion of k-word-representability and that of a graph's representation number. Also, it allows us to restrict our attention to connected graphs.

A subset of word-representable graphs is the set of permutationally representable graphs, which happens to be exactly the class of comparability graphs. The significance of this class of graphs is that in a word-representable graph, the neighbourhood of each vertex is permutationally representable, while the converse to this statement does not hold.

3.8 Exercises

1. Build the *alternation matrix*, similar to that in Table 3.1 related to Example 3.0.4, for the word $w = 243251415$. Based on the matrix, draw the graph corresponding to the word w.

2. Label the tree in Figure 3.11 in any way of your choice. Then represent this tree following the approach in Section 3.1.

3. Represent the 6-cycle C_6 following the approach presented in Section 3.2.

4. Suppose that a graph G is represented by the word $w = 153432531$.

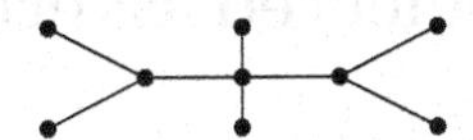

Figure 3.11: The tree for Exercise 2

(a) Following the procedure described in Section 3.3, find a (uniform) 3-word-representation of G.

(b) Draw the graph G. By 4a, G's representation number (see Definition 3.3.3) is at most 3. Show that G's representation number is actually 2.

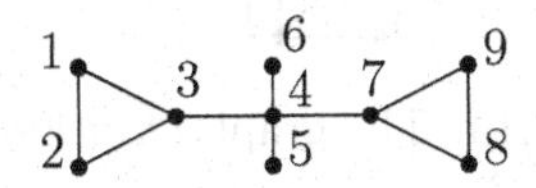

Figure 3.12: The graph G for Exercise 5

5. Represent the graph G in Figure 3.12 using two copies of each letter.
6. Consider the graph G in Figure 3.13.

   (a) Show that G is a comparability graph by providing a transitive orientation.

   (b) Label G. By 6a and Theorem 3.4.3, G can be represented by a concatenation of permutations. Find such a representation.

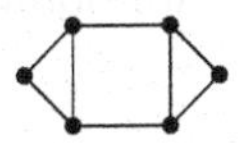

Figure 3.13: The graph G for Exercise 6

7. Repeat Exercise 6 for the cycle graph C_6 on six vertices instead of the graph G in Figure 6. Generalize your arguments for a cycle graph C_n, where n is even.
8. Let a graph G be obtained from a complete graph K_n by removing from it a matching, that is, removing a number of edges no pair of which shares a vertex. Show that G can be represented by concatenation of two permutations, and thus, by Theorem 3.4.3, G must be a comparability graph.
9. Build your own example of a non-word-representable graph on twelve vertices.

3.9 Solutions to Selected Exercises

4. For part 4a, the initial permutation $p(w) = 1542$ because the letter 3 occurs the maximum number of times and therefore is ignored. Thus, after the first step of the procedure, we obtain the word 1542153432531 representing G. One can see that the letters 1, 3 and 5 each occur three times, so that the initial permutation for the next step is 42 resulting in the desired word 421542153432531.

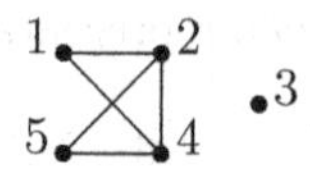

Figure 3.14: The graph G in Exercise 4

   For part 4b, the graph G is drawn in Figure 3.14. Any permutation on the set $\{1, 2, 4, 5\}$, in particular, 1245 represents the complete graph on these four vertices. We can modify this word to 12415 by inserting an extra 1 to exclude the edge 15. Finally, the word 3312415 represents G, where 33 corresponds to the isolated vertex 3. Thus, G's representation number is indeed 2.

5. Ignoring the edges 12 and 89, we can follow the procedure to represent the obtained tree described in Section 3.1 making sure that 1 stays next to 2, and 8 stays next to 9. An example of such a word representing the tree is 1232143569878965. Swapping one copy of adjacent letters 1 and 2, and one copy of adjacent 8 and 9, we obtain a word representing the graph G: 2132143569879865.

8. Suppose that G is obtained by removing k edges $a_1b_1, a_2b_2, \ldots, a_kb_k$, and $x_1, x_2, \ldots, x_{n-2k}$ are the remaining vertices. Then the word

$$a_1b_1a_2b_2\cdots a_kb_kx_1x_2\cdots x_{n-2k}b_1a_1b_2a_2\cdots b_ka_kx_1x_2\cdots x_{n-2k},$$

   obtained by concatenating two permutations of length n, represents G.

Chapter 4

Characterization of Word-Representable Graphs in Terms of Semi-transitive Orientations

Recall that, by Theorem 3.4.3, a graph is permutationally representable if and only if it is a comparability graph. The current chapter offers, based on [74, 75], a generalization of this theorem to the case of arbitrary word-representable graphs and so-called *semi-transitive orientations* (see Theorem 4.1.8). These acyclic orientations, being a generalization of transitive orientations related to partial orders, provide a powerful tool for studying word-representable graphs, in particular, allowing us to reformulate word-representability problems in pure graph-theoretical terms (via certain orientations of graphs).

As noted in [74], other orientations have been defined in the literature in order to generalize comparability graphs, such as *perfectly orderable graphs* and their subclasses including chordal graphs. We refer the reader to [27] for the definition of perfectly orderable graphs and quote from the same reference the following characterization of these graphs: a graph is perfectly orderable if and only if it admits an acyclic orientation of its edges containing no $\bullet \rightarrow \bullet \rightarrow \bullet \leftarrow \bullet$ as an induced subgraph. Observe that a transitive orientation is an acyclic orientation avoiding $\bullet \rightarrow \bullet \rightarrow \bullet$ as an induced subgraph. Therefore, perfectly orderable graphs generalize transitively orientable graphs. However, semi-transitive orientations are independent from any of the previously known generalizations of transitive orientations.

In [75], it was observed that the proof of Lemma 2 in [74], whose role is crucial in the proof of the characterization theorem, Theorem 4.1.8, contains a mistake.

This mistake was corrected in [75].

4.1 Semi-transitive Orientations

Definition 4.1.1. A graph $G = (V, E)$ is *semi-transitive* if it admits an acyclic orientation such that for any directed path $v_1 \to v_2 \to \cdots \to v_k$ with $v_i \in V$ for all i, $1 \leq i \leq k$, either

- there is no edge $v_1 \to v_k$, or
- the edge $v_1 \to v_k$ is present and there are edges $v_i \to v_j$ for all $1 \leq i < j \leq k$. In other words, in this case, the (acyclic) subgraph induced by the vertices $v_1, \ldots, v_k$ is transitive (with the unique source v_1 and the unique sink v_k).

We call such an orientation *semi-transitive.*

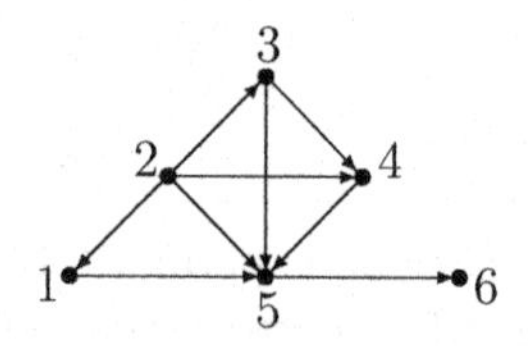

Figure 4.1: An example of a semi-transitive orientation

Example 4.1.2. The orientation of the graph in Figure 4.1 is semi-transitive , and thus the underlying (non-directed) graph is semi-transitive. Indeed, it is straightforward to see that the digraph is acyclic and we only need to worry about directed paths on three or more edges. There are six such paths:

- $2 \to 1 \to 5 \to 6$. It is nothing to worry about, since $2 \to 6$ is not an edge.
- $2 \to 3 \to 5 \to 6$. It is nothing to worry about, since $2 \to 6$ is not an edge.
- $2 \to 3 \to 4 \to 5$. Since $2 \to 5$ is an edge, we need to make sure that $2 \to 4$ and $3 \to 5$ are also edges, which is the case.
- $2 \to 3 \to 4 \to 5 \to 6$. It is nothing to worry about, since $2 \to 6$ is not an edge.
- $2 \to 4 \to 5 \to 6$. It is nothing to worry about, since $2 \to 6$ is not an edge.
- $3 \to 4 \to 5 \to 6$. It is nothing to worry about, since $3 \to 6$ is not an edge.

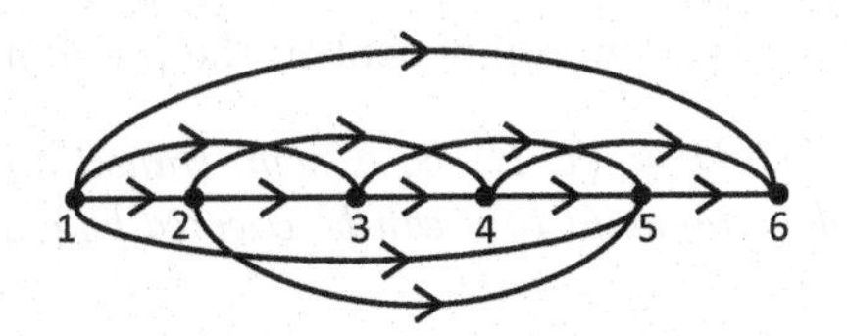

Figure 4.2: An example of a shortcut

Clearly, all transitive (that is, comparability) graphs are semi-transitive, and thus semi-transitive orientations are a generalization of transitive orientations.

Remark 4.1.3. Semi-transitive orientations are defined in [74, 75] in terms of *shortcuts* as follows. A *semi-cycle* is the directed acyclic graph obtained by reversing the direction of one edge of a directed cycle. An acyclic digraph is a shortcut if it is induced by the vertices of a semi-cycle and contains a pair of non-adjacent vertices. Thus, a digraph on the vertex set $\{v_1, \ldots, v_k\}$ is a shortcut if it contains a directed path $v_1 \to v_2 \to \cdots \to v_k$, the edge $v_1 \to v_k$, and it is missing an edge $v_i \to v_j$ for some $1 \leq i < j \leq k$; in particular, we must have $k \geq 4$, so that any shortcut is on at least four vertices. Sometimes, we slightly abuse our terminology and refer to the "shortcutting edge" $v_1 \to v_k$ as a shortcut. See Figure 4.2 for an example of a shortcut (there, the edges $1 \to 4$, $2 \to 6$ and $3 \to 6$ are missing). An orientation of a graph is semi-transitive if it is acyclic and contains no shortcuts. Clearly, this definition is just another way to introduce the notion of semi-transitive orientations presented in Definition 4.1.1.

Definition 4.1.4. A *linear extension* (also known as a *topological order* or *topsort*) of an acyclic digraph D is a permutation π of the vertices that obeys the edges, that is, for each edge $u \to v$ in D, u precedes v in π.

Example 4.1.5. All linear extensions of the digraph in Figure 4.1 are 231456, 213456, and 234156.

Recall from Definition 3.0.9 that for a word w, $G(w)$ denotes the graph represented by w. Also, recall Proposition 3.6.9 essentially saying that a word-representant w induces an orientation on the graph it represents; the respective oriented graph is denoted by $Alt_1(w)$ appearing in Definition 3.6.5. That is, for alternating letters x and y in w, we orient the edge $x \to y$ if the leftmost occurrence of x is to the left of the leftmost occurrence of y in w; we orient the edge $y \to x$ otherwise. Thus, we can extend our definition of a word-representable graph to a word-representable directed graph by paying attention to which letter in a pair of letters occurs first.

Definition 4.1.6. A word w *covers* a set of non-edges A in a graph if each non-edge in A is also a non-edge in $G(w)$ (in the undirected case) or in $Alt_1(w)$ (in the directed case).

The following lemma is a key component in the proof of Theorem 4.1.8.

Lemma 4.1.7. *([75]) Let $D = (V, E)$ be a semi-transitively oriented graph and $v \in V$. Then the non-edges incident to v can be covered by a 2-uniform word, which respects all edges in E.*

Proof. We let $I(v) = \{u : u \to v\}$ be the set of all *in-neighbours* of v, and $O(v) = \{u : v \to u\}$ be the set of all *out-neighbours* of v. Also, let $A(v)$ be the set of v's non-neighbouring vertices that can reach v, and $B(v)$ be the set of v's non-neighbouring vertices that can be reached from v. Finally, let $T(v) = V - (\{v\} \cup I(v) \cup O(v) \cup A(v) \cup B(v))$ be the set of vertices that are independent of v. Note that the sets $I(v)$, $O(v)$, $A(v)$, $B(v)$ and $T(v)$ can be empty, but they must be pairwise disjoint.

Let each of the letters A, B, I, O and T represent a 1-uniform word (a permutation) that is consistent with some topological order of the respective set $A(v)$, $B(v)$, $I(v)$, $O(v)$ and $T(v)$.

We now consider the 2-uniform word w given by

$$w = A\ I\ T\ A\ v\ O\ I\ v\ B\ T\ O\ B\ .$$

Observe that there are no edges from $A(v)$ to $O(v) \cup B(v)$, or from $I(v)$ to $B(v)$, since that would induce a shortcut, and these are consistent with the subsequences $AAOBOB$ and $IIBB$ in w, respectively. Also, there are no edges in the other direction, since the digraph is acyclic, which is consistent with A and I being the leftmost letters in w.

Moreover, by definition of $I(v)$ and $O(v)$, vertices in $T(v)$ are not reachable from $O(v) \cup B(v)$, and vertices in $A(v) \cup I(v)$ are not reachable from $T(v)$, which is consistent with the leftmost occurrence of T being to the left of the leftmost occurrences of O and B, and that occurrence of T being to the right of the leftmost occurrences of A and I in w, respectively.

Claim 1. *The word w is consistent with D, i.e. the vertices of each edge in D alternate in w.*

Proof. The edges in D are of three types:

(a) edges within each set $I(v), O(v), A(v), B(v)$ and $T(v)$,

(b) edges between v and $I(v) \cup O(v)$, and

(c) edges from $I(v) \cup A(v)$ to $T(v)$, from $T(v)$ to $O(v) \cup B(v)$, from $I(v)$ to $O(v)$, from $A(v)$ to $I(v)$, and from $O(v)$ to $B(v)$.

The edges of type (a) are automatically satisfied in w, since w is constructed using exactly two copies of each letter in $\{I, O, A, B, T\}$ and two copies of v. Also, (b) follows either from the subsequence $IvIv$ or the subsequence $vOvO$ of w, respectively. Finally, (c) follows from the following subsequences of w, respectvely: $ITIT$ and $ATAT$; $TOTO$ and $TBTB$; $IOIO$, $AIAI$ and $OBOB$. □

Claim 2. *The word w covers all non-edges incident to v.*

Proof. The vertices that v is not adjacent to are exactly those in $A(v)$, $B(v)$ and $T(v)$. It is easily verifiable by inspection that the letters corresponding to each of these do not alternate with v in w. □

The lemma is proved by Claims 1 and 2. □

The following theorem is a key result on word-representable graphs that provides a useful characterization of these graphs in terms of certain orientations.

Theorem 4.1.8. *([75]) A graph G is word-representable if and only if it admits a semi-transitive orientation.*

Proof. We start with the forward direction. Let w be a word-representant for G. By Theorem 3.3.1, we can assume that w is t-uniform for some t. We turn G into a digraph D by directing an edge (x, y) in G from x to y if in w, the first (leftmost) occurrence of x is before that of y, which is denoted $x < y$. Clearly, D is acyclic. Let us show that the orientation of D is semi-transitive.

Indeed, assume that $x_1 \to x_m$ is an edge in D and there is a directed path $x_1 \to x_2 \to \cdots \to x_m$ in D. Further, let π be obtained from w by substituting the ith copy of a letter x by x^i. Then in π we must have $x_1^i < x_2^i < \cdots < x_m^i$ for every i, $1 \leq i \leq t$. Since $x_1 \to x_m$ is an edge in D, we have that $x_m^i < x_1^{i+1}$. But then for every $j < k$ and i, there must be $x_j^i < x_k^i < x_j^{i+1}$, that is, $x_j \to x_k$ is an edge in D. So, D is semi-transitively oriented.

For the other direction, let K be a maximum clique of G and $\kappa(G)$ be its size. Denote by D a semi-transitive orientation of the graph G. Let w_v be a 2-uniform word that covers all the non-edges in G incident to a vertex $v \in V \setminus K$; w_v exists by Lemma 4.1.7. Concatenating $n - \kappa(G)$ such words w_v induces a word w that covers all non-edges in G and preserves all edges. This follows from the fact that every non-edge has at least one endpoint outside of K. Thus, G is represented by w. Moreover, w is $2(n - \kappa(G))$-uniform. □

In the rest of the chapter, we discuss a number of corollaries to Theorem 4.1.8.

4.2 Graph's Representation Number and Complexity Issues

It is an easy fact, recorded in Theorem 5.1.1, that any complete graph is 1-word-representable. The following theorem provides an upper bound on the representation number for all other word-representable graphs.

Theorem 4.2.1. *([75]) Each non-complete word-representable graph G is $2(n-\kappa(G))$-word-representable, where $\kappa(G)$ is the size of the maximum clique in G.*

Proof. The statement follows from the proof of Theorem 4.1.8. □

A direct corollary to Theorem 4.2.1 is the following result.

Theorem 4.2.2. *([74]) The recognition problem for word-representable graphs is in NP.*

Proof. Given a graph G with the vertex set $V = \{1, 2, \ldots, n\}$, and a word w of length $2n^2$, where each letter in V occurs $2n$ times, it is polynomially verifiable whether w represents G. Indeed, one needs $O(n^2)$ passes through w to check alternation properties of all pairs of letters (there are $\binom{n}{2}$ such pairs). By Theorem 4.2.1, we do not need to consider words of length larger than $2n^2$ (in fact, of length larger than $2n(n-\kappa(G))$) in order to try to represent G, and thus the decision problem of whether G is word-representable or not is solvable in $n^{O(n)}$-time. □

Remark 4.2.3. An alternative proof of Theorem 4.2.2, which is based on Theorem 4.1.8, was offered by Magnús M. Halldórsson. This proof works as follows. Checking that a given directed graph G is acyclic is a polynomially solvable problem. Indeed, it is well known that the entry (i, j) of the kth power of the adjacency matrix of G records the number of walks of length k in G from the vertex i to the vertex j. Thus, if G has n vertices, then we need to make sure that the diagonal entries are all 0 in all powers, up to the nth power, of the adjacency matrix of G. Therefore, it remains to show that it is polynomially solvable to check that G is shortcut-free. Let $u \to v$ be an edge in G. Consider the induced subgraph $H_{u\to v}$ consisting of vertices "in between" u and v, that is, the vertex set of $H_{u\to v}$ is

$$\{x \mid \text{there exist directed paths from } u \text{ to } x \text{ and from } x \text{ to } v\}.$$

It is not so difficult to prove that $u \to v$ is not a shortcut (that is, is not a shortcutting edge) if and only if $H_{u\to v}$ is transitive. Now, we can use the well known fact that finding out whether there exists a directed path from one vertex to another in a directed graph is polynomially solvable, and thus it is polynomially solvable to determine $H_{u \to v}$ (one needs to go through n vertices and check the existence of two paths for each vertex). Finally, checking transitivity is also polynomially solvable, which is not difficult to see.

By Theorem 4.2.1, the representation number $\mathcal{R}(G) \leq 2(n - \kappa(G))$ for any word-representable graph G on n vertices. A natural question is how close this upper bound is to the best (smallest) possible. The next subsection will show that the upper bound is asymptotically optimal.

4.2.1 Graphs with High Representation Number

We now show that there are graphs on n vertices with representation number $\lfloor \frac{n}{2} \rfloor$, matching the upper bound in Theorem 4.2.1 within a factor of 4.

Definition 4.2.4. A *crown graph* $H_{k,k}$ is a graph obtained from the complete bipartite graph $K_{k,k}$ by removing a perfect matching. Examples of crown graphs are given in Figure 7.4. Denote by G_k the graph obtained from a crown graph $H_{k,k}$ by adding an all-adjacent vertex.

Example 4.2.5. See Figure 4.3 for graphs involved in defining G_3.

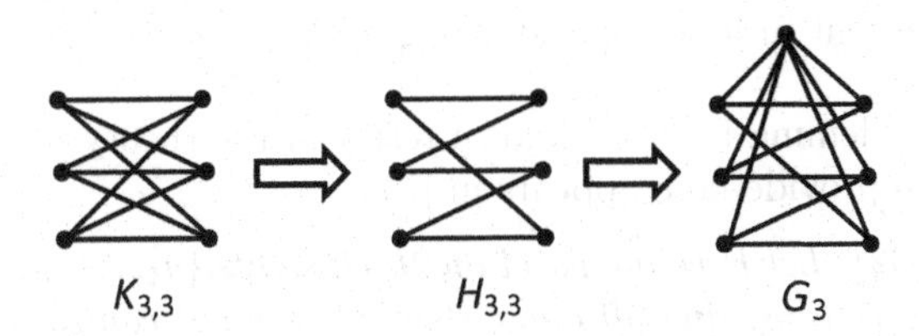

Figure 4.3: Graphs involved in defining the graph G_3

Theorem 4.2.6. *([74]) The graph G_k on $2k + 1$ vertices belongs to $\mathcal{R}_k$. That is, G_k is k-word-representable but not $(k-1)$-word-representable.*

The proof is based on three lemmas presented below. The following lemma is a refinement of Theorem 3.4.7 and it is proved in essentially the same way. Before proceeding, one should review the notion of permutationally k-representability in Definition 3.4.1.

Lemma 4.2.7. *([74]) Let H be a graph and G be the graph obtained from H by adding an all-adjacent vertex. Then G is k-word-representable if and only if H is permutationally k-representable.*

Proof. Let 0 be the letter corresponding to the all-adjacent vertex. Then every other letter of a word w representing G must appear exactly once between any two consecutive 0s. By Proposition 3.2.7, we may assume that w starts with 0. Then the word $w \setminus \{0\}$, formed by deleting all occurrences of 0 from w, is a word-representant for H obtained by writing k permutations next to each other. Conversely, if w' permutationally k-represents H, then we insert 0 in front of each permutation to obtain a (permutational) k-word-representation of G. □

Recall that the *dimension* of a poset is the minimum number of linear orders such that their intersection induces this poset (see Definition B.1.9).

Lemma 4.2.8. *([74]) A comparability graph H is permutationally k-representable if and only if the poset induced by this graph has dimension at most k.*

Proof. Let w be a word permutationally k-representing H. Each permutation in w can be considered as a linear order where $a < b$ if a occurs before b in the permutation (and vice versa). We want to show that the comparability graph of the poset induced by the intersection of these linear orders coincides with H.

Two vertices a and b are adjacent in H if and only if their letters alternate in the word. So, they must be in the same order in each permutation, that is, either $a < b$ in every linear order or $b < a$ in every linear order. But this means that a and b are comparable in the poset induced by the intersection of the linear orders, that is, a and b are adjacent in its comparability graph. □

The following lemma is a well known result in the theory of partially ordered sets. The proof we provide here appears in [74].

Lemma 4.2.9. *([74]) Let P be the poset on $2k$ elements $\{a_1, \ldots, a_k, b_1, \ldots, b_k\}$ such that $a_i < b_j$ for every $i \neq j$ and all other elements are not comparable. Then P has dimension k.*

Proof. Assume that P is the intersection of t linear orders. Since a_i and b_i are not comparable for each i, there must exist a linear order where $b_i < a_i$. If we have in some linear order both $b_i < a_i$ and $b_j < a_j$ for $i \neq j$, then either $a_i < a_j$ or $a_j < a_i$ in it. In the former case, we have that $b_i < a_j$, while in the latter case $b_j < a_i$. But each of these inequalities contradicts the definition of the poset. Therefore, $t \geq k$.

In order to show that $t = k$, we can consider a linear order

$$a_1 < a_2 < \cdots < a_{k-1} < b_k < a_k < b_{k-1} < \cdots < b_2 < b_1$$

together with all linear orders obtained from this order by simultaneous exchange of a_k and b_k with a_m and b_m respectively ($m = 1, \ldots, k-1$). It can be verified that the intersection of these k linear orders coincides with our poset. □

Now, we can prove Theorem 4.2.6 as is done in [74]. Since the crown graph $H_{k,k}$ is a comparability graph of the poset P, we deduce from Lemmas 4.2.9 and 4.2.8 that $H_{k,k}$ is permutationally k-representable but not permutationally $(k-1)$-representable. Then by Lemma 4.2.7 we have that G_k is k-word-representable but not $(k-1)$-word-representable. Theorem 4.2.6 is proved. □

4.2.2 Three Complexity Statements

From Lemmas 4.2.7 and 4.2.8, we see that deciding the complexity of determining the graph's representation number is as hard as determining the dimension of a poset. Yannakakis [135] showed that for any k, $3 \leq k \leq \lceil n/2 \rceil$, it is NP-hard to determine whether a given poset has dimension k. Thus, we have the following theorem.

Theorem 4.2.10. *([74]) Deciding whether a given graph is k-word-representable, for any fixed k, $3 \leq k \leq \lceil n/2 \rceil$, is NP-complete.*

Hegde and Jain [78] showed that it is NP-hard to approximate the dimension of a poset within almost a square root factor. We therefore obtain the same hardness for the graph's representation number. However, it is noted in [75] that based on a recent paper [33] by Chalermsook et al., "$n^{1/2-\epsilon}$" can be replaced by "$n^{1-\epsilon}$" in that result, which is recorded in the following theorem.

Theorem 4.2.11. *([75]) Approximating the representation number within a factor of $n^{1-\epsilon}$ is NP-hard, for any $\epsilon > 0$. That is, for every constant $\epsilon > 0$, it is an NP-complete problem to decide whether a given word-representable graph has representation number at most n^{ϵ}, or it has representation number greater than n^{ϵ}.*

In contrast with these hardness results, the case $k = 2$ turns out to be easier and admits a succinct characterization (see Section 5.1). Given the NP-hardness result in Theorem 4.2.10, a similarly "nice" characterization of k-word-representable graphs for $k \geq 3$ is unlikely.

Suppose that P is a poset and x and y are two of its elements. We say that x *covers* y if $x > y$ and there is no element z in P such that $x > z > y$ (see Definition B.1.7).

Definition 4.2.12. The *cover graph* G_P of a poset P has P's elements as its vertices, and $\{x, y\}$ is an edge in G_P if and only if either x covers y, or vice versa. The *diagram* of P, sometimes called a *Hasse diagram* or *order diagram*, is a drawing of the cover graph of G in the plane with x being higher than y whenever x covers y in P.

Example 4.2.13. The three-dimensional cube in Figure 5.6 is an example of a cover graph.

It was pointed to us by Vincent Limouzy [104] in 2014 that semi-transitive orientations of triangle-free graphs are exactly the 2-*good orientations* considered in [116] by Pretzel (we refer to that paper for the definition of a k-good orientation). Thus, by Proposition 1 in [116] we have the following reformulation of Pretzel's result in our language.

Theorem 4.2.14. *The class of triangle-free word-representable graphs is exactly the class of cover graphs of posets.*

It was further pointed to us by Limouzy [104] in 2014, that it is an NP-complete problem to recognize the class of cover graphs of posets [28]. This implies the following theorem, which is a key complexity result on word-representable graphs in this book.

Theorem 4.2.15. *It is an NP-complete problem to recognize whether a given graph is word-representable.*

4.2.3 A Summary of Complexity Results

Even though Theorem 3.4.14 shows that the Maximum Clique problem is polynomially solvable on word-representable graphs, Theorem 4.3.3 below implies that many classical optimization problems are NP-hard on these graphs, which is recorded in the following theorem.

Theorem 4.2.16. *([74]) The following optimization problems are NP-hard on word-representable graphs:*

- Dominating Set,
- Vertex Colouring,
- Clique Covering *and*
- Maximum Independent Set.

Our knowledge of complexity issues related to word-representable graphs is manifested in Theorems 3.4.14, 4.2.10, 4.2.11, 4.2.16 and 4.2.15, and we give a summary in Table 4.1.

In the rest of the section we give definitions of relevant notions and the problems appearing in Theorem 4.2.16.

problem	complexity
deciding whether a given graph is word-representable	NP-complete
approximating the graph representation number within a factor of $n^{1-\epsilon}$ for any $\epsilon > 0$	NP-hard
Clique Covering	NP-hard
deciding whether a given graph is k-word-representable for any fixed k, $3 \leq k \leq \lceil n/2 \rceil$	NP-complete
Dominating Set	NP-hard
Vertex Colouring	NP-hard
Maximum Clique	polynomially solvable
Maximum Independent Set	NP-hard

Table 4.1: Known complexities for problems on word-representable graphs

Definition 4.2.17. An *independent set*, or *stable set*, is a set of vertices in a graph, no two of which are adjacent. An independent set S is *maximal* if no other independent set properly contains S and is *maximum* if no independent set has cardinality greater than $|S|$.

Example 4.2.18. An example of a maximum independent set in the graph in Figure 3.4 is vertices 3 and 7.

Definition 4.2.19. Given a graph G, the *Maximum Independent Set problem* is the problem of finding a maximum independent set in G.

Definition 4.2.20. A *dominating set* for a graph $G = (V, E)$ is a subset D of V such that every vertex not in D is adjacent to at least one member of D. The *domination number* $\gamma(G)$ is the number of vertices in a smallest dominating set for G.

Example 4.2.21. For the graph co-(T_2) in Figure 3.4, $\gamma(\text{co-}(T_2)) = 2$ because there are no all-adjacent vertices in the graph, while, say, the vertices 1 and 5 form a dominating set.

Definition 4.2.22. Given a graph G, the *Dominating Set problem* is the problem of finding a minimum dominating set in G.

Definition 4.2.23. A *clique covering* of a graph G is a set of cliques such that every vertex of G is a member of at least one clique. The *Clique Covering problem* is the problem of determining whether the vertices of a given graph can be partitioned into k cliques.

Definition 4.2.24. The *Vertex Colouring problem* is the problem of colouring the vertices of a given graph by using the smallest number of colours so that adjacent vertices receive different colours.

4.3 Chromatic Number and Graphs Representable by Words

Recall the notions of a k-colourable graph and the chromatic number from Definition 2.2.23.

A 1-colourable graph is a graph with no edges.

Theorem 4.3.1. *Any* 1*-colourable graph* G *is* 2*-word-representable.*

Proof. Suppose the vertex set of G is $\{1,\ldots,n\}$. Then it is easy to see that the 2-uniform word $12\cdots nn(n-1)\cdots 1$ represents G. □

The class of 2-colourable graphs is exactly the class of bipartite graphs.

Theorem 4.3.2. *Any* 2*-colourable graph is permutationally representable.*

Proof. Each bipartite graph is a comparability graph (orienting all edges, if any, from one part to the other gives a transitive orientation), and the result now follows from Theorem 3.4.3. □

Given a 3-colourable graph, direct its edges from the first colour group through the second to the third group. It is easy to see that we obtain a semi-transitive digraph as there will be no shortcuts in it, and Theorem 4.1.8 can be applied to see that any 3-colourable graph is word-representable. Note though that the upper bound of $2n-4$ for the graph representation number given by Theorem 4.2.1 can be improved for 3-colourable graphs. We record the improvement in the next theorem, which also provides an alternative (self-contained) proof of the fact that any 3-colourable graph is word-representable.

Theorem 4.3.3. *([75])* 3*-colourable graphs are* $2\lfloor 2n/3\rfloor$*-word-representable.*

Proof. Let G be a 3-colourable graph.

Suppose that the vertices of G are partitioned into three independent sets A, B and C. We may assume that the set B has the maximum cardinality among these three sets. Direct the edges of G from A to $B\cup C$ and from B to C, and denote the obtained orientation by D.

Let $v\in A$. Denote by N_v^B (resp., N_v^C) an arbitrary permutation over the set of neighbours of v in B (resp., in C) and by $\overline{N_v^B}$ (resp., $\overline{N_v^C}$) an arbitrary permutation over the remaining vertices of B (resp., C). Also, let A_v be an arbitrary permutation over the set $A\backslash\{v\}$. Recall that for a permutation π, $r(\pi)$ is the permutation written in the reverse order.

Consider the 2-uniform word

$$w_v^A = A_v\ \overline{N_v^B}\ v\ N_v^B\ N_v^C\ v\ \overline{N_v^C}\ r(A_v)\ r(N_v^B)\ r(\overline{N_v^B})\ r(\overline{N_v^C})\ r(N_v^C).$$

Note that w_v^A covers all non-edges lying inside B and C and also all non-edges incident to v. Indeed, the graph H induced by w_v^A is obtained from the complete tripartite graph with the partition A, B, C by removing all edges connecting v with $\overline{N_v^B} \cup \overline{N_v^C}$. Since the initial permutation $p(w_v^A) = A_v\ \overline{N_v^B}\ v\ N_v^B\ N_v^C\ \overline{N_v^C}$ and no directed path can go from v to $\overline{N_v^B}$, it is a topological sort of D.

Similarly, for $v \in C$ consider the 2-uniform word

$$w_v^C = N_v^A\ \overline{N_v^A}\ \overline{N_v^B}\ N_v^B\ C_v\ r(\overline{N_v^A})\ v\ r(N_v^A)\ r(N_v^B)\ v\ r(\overline{N_v^B})\ r(C_v),$$

where C_v, $\overline{N_v^A}$, N_v^A, $\overline{N_v^B}$ and N_v^B are defined by analogy with the respective sets above. Using similar arguments, one can show that w_v^C covers all non-edges lying inside A and B and all non-edges incident to v.

Concatenating all these words, we obtain a $2k$-uniform word w representing G, where $k = |A \cup C|$. Since B has the maximum cardinality, $k \leq \lfloor 2n/3 \rfloor$. □

Theorem 4.3.3 implies a number of results on word-representability proved by other means, including those of outerplanar graphs, subdivision graphs and prisms (see Theorems 5.1.6, 5.2.8 and 5.2.15 in Sections 5.1 and 5.2). Theorems 4.3.6, 4.3.7 and 4.3.9 below are also corollaries of Theorem 4.3.3.

Definition 4.3.4. A *k-degenerate graph* is an undirected graph in which every subgraph has a vertex of degree at most k, that is, some vertex in the subgraph is incident to k or fewer of the subgraph's edges.

Example 4.3.5. Every planar graph has a vertex of degree 5 or less; therefore, every planar graph is 5-degenerate.

It can be proved by a simple induction on the number of vertices, which is very similar to the proof of the *six-colour theorem for planar graphs*, that a k-degenerate graph has chromatic number at most $k+1$. From this fact and Theorem 4.3.3, we have the truth of the following theorem.

Theorem 4.3.6. *([74]) Any 2-degenerate graph is word-representable.*

The following theorem is essentially equivalent, via Brooks' Theorem (see Theorem 2.2.24), to the fact that all 3-colourable graphs are word-representable.

Theorem 4.3.7. *([74]) Graphs of maximum degree 3, i.e. subcubic graphs, are word-representable. In particular, 3-regular graphs are word-representable.*

k-colourable graphs	**word-representability**
$k = 1$	2-word-representable
$k = 2$	permutationally representable
$k = 3$	word-representable
$k \geq 4$	can be either word-representable or not

Table 4.2: An overview of word-representability of k-colourable graphs

Proof. This is an immediate consequence of Theorems 4.3.3 and 2.2.24 taking into account that a complete graph K_n is represented by a single permutation of length n, while any cycle graph is 3-colourable and thus is word-representable (actually, 2-word-representable; see Section 3.2). □

In 1959, Herbert Grötzch published the following theorem whose proof was simplified in [130].

Theorem 4.3.8. *(Grötzch's theorem [130]) Every triangle-free planar graph can be coloured with only three colours.*

Theorem 4.3.9. *([74]) Triangle-free planar graphs are word-representable.*

Proof. This is an immediate consequence of Theorems 4.3.3 and 4.3.8. □

Remark 4.3.10. Note that it is essential in Theorem 4.3.9 that we deal with planar graphs: in Section 4.4, we will see examples of triangle-free (non-planar) graphs that are not word-representable.

It turns out that we cannot go for higher chromatic numbers — the examples in Figure 3.6 show that 4-colourable graphs can be non-word-representable. Thus, a short summary of the situation with k-colourable graphs can be presented as in Table 4.2. However, one can obtain more results in this direction, for example, in terms of the girth of a graph.

Definition 4.3.11. The *girth* of a graph is the length of a shortest cycle contained in the graph. If the graph does not contain any cycles (that is, it is an acyclic graph), its girth is defined to be infinity.

Example 4.3.12. A 4-cycle C_4 (square) has girth 4, while the graph in Figure 3.8 has girth 3. A graph with girth 4 or more is *triangle-free.*

Theorem 4.3.13. *([74]) Let G be a graph whose girth is greater than its chromatic number. Then, G is a word-representable graph.*

Proof. Suppose that G is coloured using the colours $\{1,\ldots,\chi(G)\}$. Direct the edges of the graph from smaller to larger colours. There is no directed path with more than $\chi(G)-1$ edges, but since G contains no cycle of $\chi(G)$ or fewer edges, there can be no shortcuts. Hence, G is semi-transitively oriented and it must be word-representable by Theorem 4.1.8. □

The Petersen graph (presented in Figure 3.1) is 3-colourable and thus, by Theorem 4.3.3, it is word-representable. In Theorem 5.2.1 we will refine this result by showing that this graph is 3-word-representable but not 2-word-representable.

4.4 Triangle-Free Non-word-representable Graphs

The following theorem shows how to construct an infinite family of triangle-free non-word-representable graphs, thus enlarging the class of non-word representable graphs constructed in Subsection 3.5.1. To prove this theorem, we need the following well known fact first established by Paul Erdős [50] using the probabilistic method in 1959 (recall the notion of girth in Definition 4.3.11): for any positive integers k and ℓ, there exists a graph with girth at least k and chromatic number at least ℓ.

Theorem 4.4.1. *([74]) There exist triangle-free non-word-representable graphs.*

Proof. Let M be a 4-chromatic graph with girth at least 10. For every path P of length 3 in M add to M the edge e_P connecting its ends. Denote the obtained graph by G. Let us show that G is a triangle-free non-word-representable graph.

If G contains a triangle on the vertices u, v and w, then M contains three paths P_{uv}, P_{uw} and P_{vw} of lengths 1 or 3 connecting these vertices (we have length 1 if the respective vertices are adjacent in M, and 3 if the vertices are not connected in M). Let T be a graph spanned by these three paths. Since T has at most nine edges and the girth of M is at least 10, T is a tree. Clearly, it cannot be a path. So, it is a subdivision (see Definition 5.2.4) of $K_{1,3}$ with the leaves u, v and w. But then at least one of the paths P_{uv}, P_{uw} or P_{vw} must have even length, a contradiction.

So, G is triangle-free. Assume that G has a semi-transitive orientation. Then it induces a semi-transitive orientation on M. Since M is 4-chromatic, each of its acyclic orientation must contain a directed path P of length at least 3 (otherwise, we would be able to subdivide the vertices of M into three groups with numbers 1, 2 and 3, so that the edges are oriented from groups with smaller numbers towards groups with larger numbers, and thus M would be 3-chromatic). But then the orientation of the edge e_P in G produces either a directed 4-cycle or a shortcut, contradicting the semi-transitivity. So, G is a triangle-free non-word-representable graph. □

4.5 Non-word-representable Graphs of Maximum Degree 4

It was an open question for several years whether there exist non-word-representable graphs of maximum degree 4. The following theorem answers this question in the affirmative by providing a construction of infinitely many such graphs.

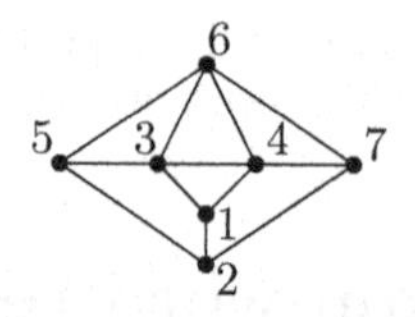

Figure 4.4: A non-word-representable graph of maximum degree 4

Before dealing with the next theorem, note that for any of the partial orientations of the 3- or 4-cycles given in Figure 4.5, there is a unique way of completing the orientation, also shown in Figure 4.5, so that oriented cycles and shortcuts are avoided. This stays true in the context of triangulated 4-cycles, because such graphs are different from K_4, admitting as they do an alternative semi-transitive (in fact, transitive) orientation completion.

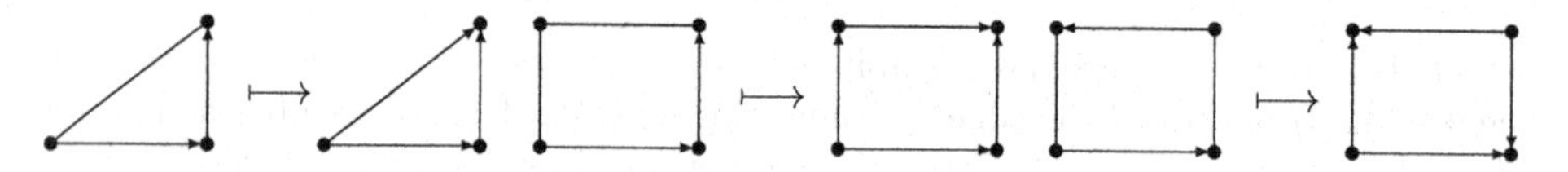

Figure 4.5: Unique way of completing in a semi-transitive way partial orientations of a 3-cycle or a 4-cycle

In the proof of Theorem 4.5.1 we use the following terminology introduced in [3]. *Complete XYW(Z)* refers to completing the orientations on a cycle $XYW(Z)$ according to the respective cases in Figure 4.5. Instances in which it is not possible to uniquely determine orientations of any additional edges in a partially oriented graph are referred to as *Branching XY*. Here, one picks a new, still non-oriented edge XY of the graph and assigns the orientation $X \rightarrow Y$, while, at the same time, one makes a copy of the graph with its partial orientations and assigns orientation $Y \rightarrow X$ to the edge XY. The new copy is named and examined later on. Our terminology and relevant abbreviations are summarized in Table 4.3.

Theorem 4.5.1. *([40]) There exist (infinitely many) non-word-representable graphs of maximum degree* 4.

Abbreviation	**Operation**
B	Branch
NC	Obtain a new partially oriented copy
C	Complete
MC	Move to a copy
S	Obtain a shortcut

Table 4.3: List of used operations and their abbreviations

Proof. We prove the theorem by showing that the graph in Figure 4.4 is not semi-transitive and thus it is not word-representable. Our proof is different from that given in [40]. We can then attach paths of arbitrary lengths to vertices 1, 2, 5 and 7 to obtain infinitely many non-word-representable graphs of maximum degree 4.

Name A the first copy of the graph in Figure 4.4 with the single edge orientated as $5 \to 6$, and carry out the following operations.

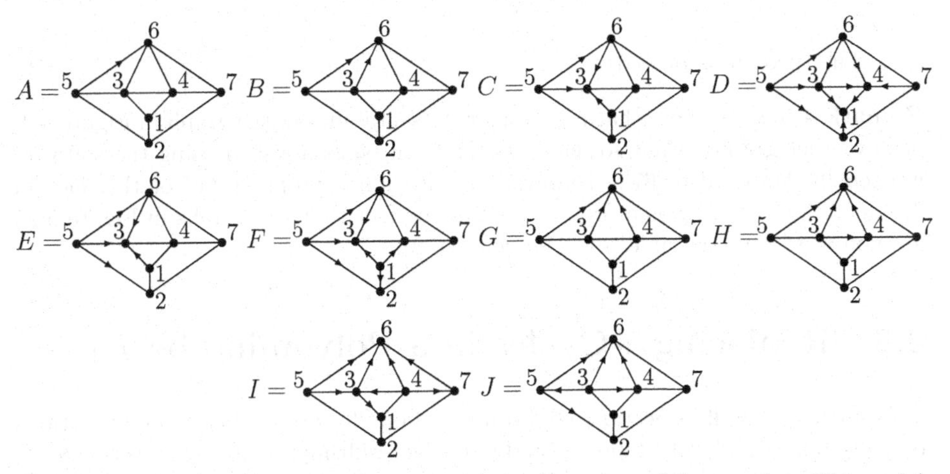

Figure 4.6: Non-existence of a semi-transitive orientation of the graph in Figure 4.4

- B 63 (NC B, see Figure 4.6), C 563, B 31 (NC C, see Figure 4.6), C 5312, C 6314, C 3564, B 47 (NC D, see Figure 4.6), C 2147, C 647, S 6347.

- MC D, C 7412, C 6347, S 5672.

- MC C, B 25 (NC E, see Figure 4.6), C 2531, C 2567, C 7634, C 1274, C 4136, S 5643.

- MC E, C 2531, B 41 (NC F, see Figure 4.6), C 4136, C 4127, C 7256, no completion of 4763 is possible, since $3 \to 4$ gives a cycle, while $4 \to 3$ gives a shortcut.
- MC F, C 4136, C 3564, C 6347, C 2147, S 5672.
- MC B, B 64 (NC G, see Figure 4.6), C 5643, C 3641, C 5312, C 2147, C 6527, S 3674.
- MC G, B 43 (NC H, see Figure 4.6), C 3465, C 4367, B13 (NC I, see Figure 4.6), C 1364, C 1472, C 2135, S 5276.
- MC I, C 1364, C 5312, C 2567, S 4721.
- MC H, C 3465, B 25 (NC J, see Figure 4.6), C 2567, C 4367, C 2741, C 1253, S 3146.
- MC J, C 3521, C 1364, C 4127, C 3476, S 5672.

The theorem is proved. □

Remark 4.5.2. We note that non-word-representability of the graph in Figure 4.4, or any other graph on not so many vertices can be established using the software written by Marc Glen [67]. However, we provide a proper proof of this fact in Theorem 4.5.1, as well as a proper proof of the fact that the graphs in Figure 5.25 are non-word-representable (see Theorem 5.6.2).

4.6 Replacing 4-Cycles in a Polyomino by K_4

This section, as well as Section 5.6, both based on [3], deals with certain operations on *polyominoes* and word-representability of the resulting graphs. This section deals with replacing 4-cycles in a polyomino by the complete graph K_4, while Section 5.6 deals with triangulations of polyominoes.

Definition 4.6.1. A *polyomino* is a plane geometric figure formed by joining one or more equal squares edge to edge.

Example 4.6.2. See the left picture in Figure 4.7 for an example of a polyomino.

Letting corners of squares in a polyomino be vertices, we can treat polyominoes as graphs. In particular, well known *grid graphs* are obtained from polyominoes in this way. Of particular interest to us in Section 5.6 are *convex polyominoes*.

Definition 4.6.3. A polyomino is said to be *column convex* if its intersection with any vertical line is convex (in other words, each column has no holes). Similarly, a polyomino is said to be *row convex* if its intersection with any horizontal line is convex. A polyomino is said to be *convex* if it is row and column convex.

Example 4.6.4. The polyomino in the left picture in Figure 4.7 is convex.

Let us consider replacement of *each* 4-cycle in a polyomino with the complete graph K_4; see Figure 4.7 for an example. It turns out that applying this operation to an *arbitrary* polyomino *always* results in a word-representable graph, unlike triangulations of polyominoes considered in Section 5.6.

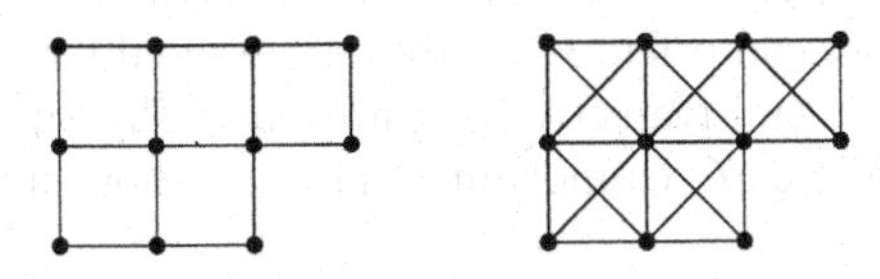

Figure 4.7: An example of replacement of 4-cycles in a polyomino by K_4

Theorem 4.6.5. *([3]) Replacing* each *4-cycle in a polyomino $\mathcal{P}$ by K_4 gives a word-representable graph $\mathcal{P}_{K_4}$.*

Proof. We begin by providing a semi-transitive orientation of the graph G obtained from a grid graph by replacing each 4-cycle by K_4, and then we discuss the case of an arbitrary polyomino.

We call a vertex in an oriented copy of G a *horizontal sink* (resp., *horizontal source*) if there are no horizontal edges coming out of (resp., coming in to) the vertex.

A semi-transitive orientation of G can now be described as follows. Make the top row of G, as well as any odd row from the top, be a sequence of alternating horizontal sources and sinks (from left to right), and all other rows a sequence of alternating horizontal sinks and sources as shown in Figure 4.8. Moreover, we orient all other edges of G downwards (see Figure 4.8). Clearly, the orientation is acyclic. Furthermore, it is easy to see that the orientation of G is semi-transitive. Indeed, starting from a vertex v and looking at paths of length 3 or more ending in vertex u, we see by inspection that there are only two possibilities:

- v and u are not connected by an edge (most common situation) thus giving no chance for a shortcut;
- there is an edge from v to u. In this case, the directed path must be of length 3 and it must cover three external edges of a K_4 oriented transitively. Thus, we do not have a shortcut in this case either.

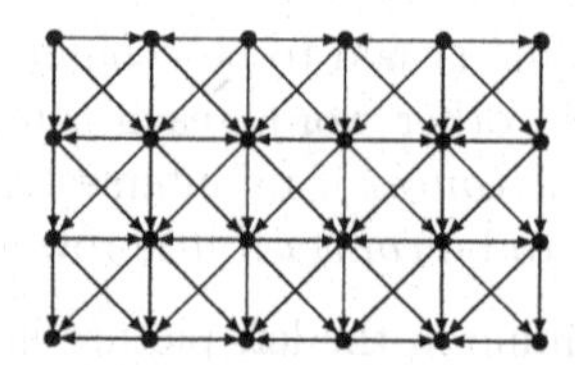

Figure 4.8: A semi-transitive orientation of G in the proof of Theorem 4.6.5

It is now clear how to semi-transitively orient $\mathcal{P}_{K_4}$ for any polyomino $\mathcal{P}$ rather than just for a grid graph. Indeed, $\mathcal{P}_{K_4}$ can be extended to a grid graph G by adding missing K_4s, then we can orient G as above, and finally remove the K_4s that were just added to obtain a semi-transitive orientation of $\mathcal{P}_{K_4}$ (it is easy to see that removing K_4s from the oriented G cannot introduce any shortcuts). □

As noted in [3], in our orientation of G in the proof of Theorem 4.6.5 (see Figure 4.8), horizontal sources in different rows are never on top of each other. A similar observation applies to horizontal sinks. If we were to eliminate this condition, thus making odd columns consist of horizontal sources and even columns of horizontal sinks keeping the vertical edges oriented in the same way, we would obtain an orientation having shortcuts (e.g., see the induced subgraph formed by the first two vertices in the top row and the second and third vertices in the second row).

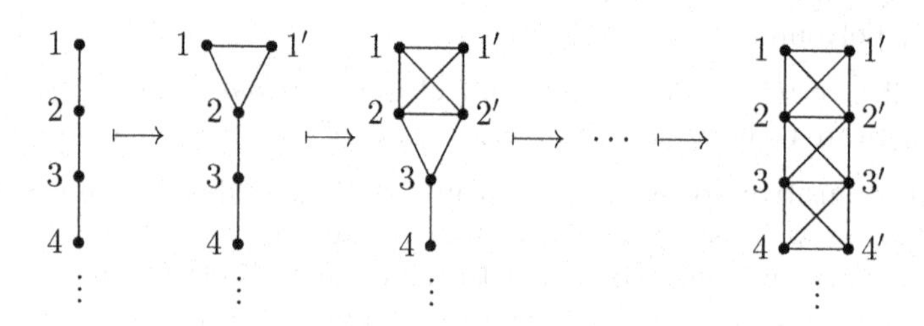

Figure 4.9: Generating $\mathcal{P}_{K_4}$ for an $n \times 2$ grid graph from a path graph on n vertices through replacing vertices by the module K_2

Another remark made in [3] is that when $\mathcal{P}$ is an $n \times 2$ (equivalently, $2 \times n$) grid graph in Theorem 4.6.5 (the respective graph $\mathcal{P}_{K_4}$ is presented schematically to the right in Figure 4.9), then we can prove this particular case of the theorem by other means. Indeed, $\mathcal{P}_{K_4}$ can be obtained in this case by substituting the vertices of a path graph (presented to the left in Figure 4.9) by *modules* K_2 (see Definition 5.4.5), as shown schematically in Figure 4.9. As we will see in Theorem 5.4.7

(in Subsection 5.4.4), replacing a vertex in a graph by a module turns a word-representable graph into a word-representable graph, showing that $\mathcal{P}_{K_4}$ in this case is word-representable because the path graph on n vertices we started with is such a graph (a particular representation of the graph is $12132435465\ldots$). Alternatively, we can directly come up with a word representing the graph $\mathcal{P}_{K_4}$, e.g. representing it by the word $11'22'11'33'22'44'33'55'\ldots$. However, none of these approaches work, at least that easily, for larger grid graphs with 4-cycles replaced by K_4, so it is essential to employ semi-transitive orientations here to prove the results.

4.7 Conclusion

In this chapter we introduced the notion of a semi-transitive orientation. This notion is to date the most powerful tool for studying graph word-representability due to the characterization theorem stating that a graph is word-representable if and only if it admits a semi-transitive orientation. As an immediate corollary to the characterization theorem we see that all 3-colourable graphs are word-representable. Another corollary is that if a graph on n vertices is word-representable then one needs at most n copies of each letter to represent the graph. We are aware of graphs requiring around $n/2$ copies of each letter to be represented. However, we do not know whether there exist graphs requiring longer word-representants.

Unfortunately, it is an NP-complete problem to recognize whether a given graph is word-representable. Moreover, many classical optimization problems are NP-hard on word-representable graphs, for example, Clique Covering, Dominating Set, Vertex Colouring and Maximum Independent Set. On the positive side though, Maximum Clique is polynomially solvable on these graphs.

4.8 Exercises

1. Determine which of the graphs in Figure 4.10 are oriented semi-transitively.

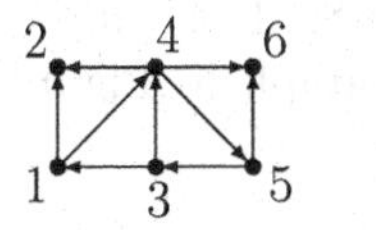

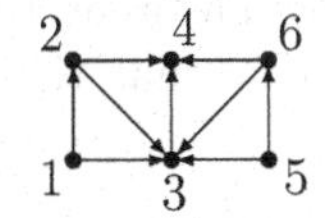

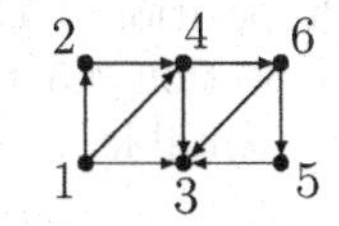

Figure 4.10: Three graphs related to Exercise 1

2. Provide two examples of linear extensions (see Definition 4.1.4) of the second graph in Figure 4.10.

3. Based on the proof of Lemma 4.1.7, provide a 2-uniform word covering the non-edges incident to the vertex 2 in the second graph in Figure 4.10.

Figure 4.11: The semi-transitively oriented graph D related to Exercise 4

4. Label the semi-transitive digraph D in Figure 4.11. Then use the proof of Theorem 4.2.1 to generate a uniform word of length 50 representing D.

5. Find explicit k-representations for the graph G_k in Theorem 4.2.6 for $k = 2, 3, 4$.

6. Show that the graph in Figure 4.12 admits a semi-transitive orientation, and thus, by Theorem 4.1.8, it is word-representable. Note that there are two alternative proofs of the word-representability fact. Namely, we can use Theorem 4.3.3 after showing that the graph in Figure 4.12 is 3-colourable, or Theorem 4.3.7 after showing that the graph is subcubic.

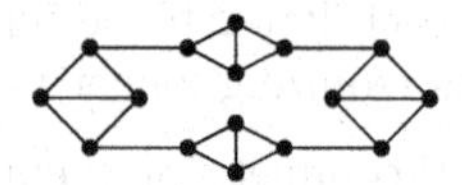

Figure 4.12: The graph for Exercise 6

7. Provide an example of a word-representable graph G discussed in Theorem 4.3.13.

8. Provide an explicit construction of a triangle-free non-word-representable graph based on the proof of Theorem 4.4.1.

9. Show that the wheel W_5 is not semi-transitively orientable, which is consistent with Proposition 3.5.3 by Theorem 4.1.8. This can be shown using the approach in Section 4.5 to show that the graph in Figure 4.4 is not semi-transitively orientable.

4.9 Solutions to Selected Exercises

1. Only the second graph in Figure 4.10 is oriented semi-transitively. Indeed, the first graph contains cycles, e.g. $3 \to 4 \to 5 \to 3$, and also shortcuts, e.g.

$5 \to 3 \to 4 \to 6$ and $5 \to 6$, while the last graph, though it is acyclic, contains shortcuts, e.g. $1 \to 4 \to 6 \to 3$ and $1 \to 3$.

3. For vertex 2 in the second graph in Figure 4.10, we have the following sets appearing in the proof of Lemma 4.1.7: $A(2) = B(2) = \emptyset$, $I(2) = \{1\}$, $O(2) = \{3,4\}$, and $T(2) = \{5,6\}$. Thus, the desired 2-uniform word representing the two non-edges $(2,5)$ and $(2,6)$ is $w = 156234125634$.

5. See Table 7.1 for representation of the graph $H_{k,k}$ for $k = 2,3,4$. The desired representations of G_k can be obtained from these by inserting the all-adjacent vertex in front of each permutation of length $2k$.

Chapter 5

Various Results on Word-Representable Graphs

For what follows, recall from Definition 3.2.3 that for a word-representable graph G, its representation number $\mathcal{R}(G)$ is the minimal k such that G is k-word-representable.

5.1 Characterization of Graphs with Representation Number at Most 2

From Definitions 3.0.3, 3.0.5, 3.2.3 and 3.3.3 we have the following characterization theorem.

Theorem 5.1.1. *For a graph G, $\mathcal{R}(G) = 1$ if and only if $G = K_n$, the complete graph on n vertices, for some n.*

While the characterization of $\mathcal{R}_1$, the class of graphs with representation number 1, is trivial, letting this number be 2 gives a more involved case. A characterization of $\mathcal{R}_2$ is given in Theorem 5.1.7. However, before stating that theorem, we will discuss various properties of graphs that are 2-word-representable that will include Theorem 5.1.6. In that discussion we follow the paper [92]. Another relevant result on 2-word-representability of *ladder graphs* is to be discussed in Subsection 5.2.3 (see Theorem 5.2.21).

Definition 5.1.2. A *cut-vertex* in a connected graph is a vertex whose deletion, together with the edges incident to it, leads to a graph with two or more connected components. A graph is 2-*connected* if it contains no cut-vertex.

Example 5.1.3. There are three cut-vertices in the graph in Figure 3.12, namely, 3, 4 and 7. On the other hand, the graph in Figure 3.13 is 2-connected.

Definition 5.1.4. A graph G is *outerplanar* if it can be drawn in the plane in such a way that no two edges meet in a point other than a common vertex and all the vertices lie in the outer-face (that is, in the unbounded face).

Example 5.1.5. An example of an outerplanar graph is given in Figure 5.1. For another example, note that any triangulation of an n-gon gives an outerplanar graph.

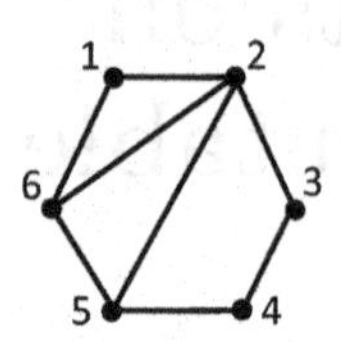

Figure 5.1: An outerplanar graph

Theorem 5.1.6. *([92]) If a graph G is outerplanar then it is 2-word-representable. Moreover, if G is also 2-connected then it can be represented by a word w such that for every edge xy incident to the outer-face, we have a factor xy or yx in w, and these factors do not overlap for different edges of the outer-face.*

Proof. We prove the theorem by induction on n, the number of vertices. For $n = 1$, the statement holds, since the only vertex, say x, is word-representable by the word xx.

If G has cut-vertices then we apply the induction hypothesis to its blocks and then connect them together using the technique of Theorem 5.4.2 which will result in 2-word-representation of G. Thus, it remains to consider the case when G is 2-connected.

Let the vertices of the outer-face of G, in clockwise direction, be labelled by $1, 2, \ldots, n$. If G has no chords, that is, G is a cycle, then we can apply the construction given in Subsection 3.2 to word-represent an n-cycle to obtain the following 2-word-representation of G satisfying the second condition of the theorem:

$$w = 1n213243\cdots n(n-1).$$

In particular, the case $n = 3$ provides the induction basis for what follows.

Suppose now that G has a chord ij where $i < j - 1$. Consider two outerplanar 2-connected graphs G_1 and G_2 with the outer-faces $12\cdots ij(j+1)\cdots n$

and $i(i+1)\cdots j$, respectively. By induction hypothesis, both of them are 2-word-representable. Moreover, using the induction hypothesis for the second condition of the theorem, and, if necessary, Propositions 3.0.14 and 3.2.7, we can assume that the words representing G_1 and G_2 are of the forms $w_1 = w_1'ij$ and $w_2 = ijw_2'$, respectively. Moreover, these words contain non-overlapping factors corresponding to all of the edges of the outer-faces. But then the word $w = w_1'w_2'$ represents G and satisfies the second condition of the theorem. □

To illustrate the steps of the proof of Theorem 5.1.6, we show how to word-represent the graph G in Figure 5.1. We begin with representing the triangle formed by the vertices 1, 2 and 6 by the word $w_1 = 162162$. Further, the triangle formed by the vertices 2, 5 and 6 can be represented by the word 265265, and thus also by the word 526526 (a 2-element cyclic shift was applied) and the word $w_2 = 625625$ (the reverse was applied). We can now use the words $w_1 = w_1'62$ and $w_2 = 62w_2'$ to represent the subgraph of G induced by the vertices 1, 2, 5 and 6 (26 is a chord there): $w_3 = w_1'w_2' = 16215625 = w_3'25$. Further, the 4-cycle formed by the vertices 2, 3, 4 and 5 can be word-represented by the word $w_4 = 25324354 = 25w_4'$ and thus, thinking of the chord 25 in G, the graph G can be word-represented by $w_3'w_4' = 162156324354$.

Recall Definition 2.2.18 of a circle graph given in Subsection 2.2.4. The following theorem was a neat observation made in [74]. It gives a characterization of $\mathcal{R}_2$, the class of graphs with representation number 2.

Theorem 5.1.7. *([74]) For a graph G different from a complete graph, $\mathcal{R}(G) = 2$ if and only if G is a circle graph.*

Proof. Suppose that G is a circle graph on n vertices that is different from a complete graph. We label the endpoints of a chord in the definition of a circle graph by a letter, which is the same as the chord's label, and we let different chords be labelled by different numbers from $\{1, 2, \ldots, n\}$. Starting with an endpoint, we read the $2n$ endpoints in clockwise direction to obtain a word w. It is straightforward to see that chords labeled x and y intersect if and only if the letters x and y alternate in w. Thus, G is 2-word-representable. Since G is different from a complete graph, G is not 1-word-representable. So, $\mathcal{R}(G) = 2$. □

To illustrate the argument in the proof of Theorem 5.1.7, we provide a circle graph in Figure 5.2 (to the left) that is 2-word-represented by the word 1233452154 obtained by reading the labels of chords (in the circle to the right) in clockwise direction.

As a direct corollary to Theorems 5.1.6 and 5.1.7 we have the following statement.

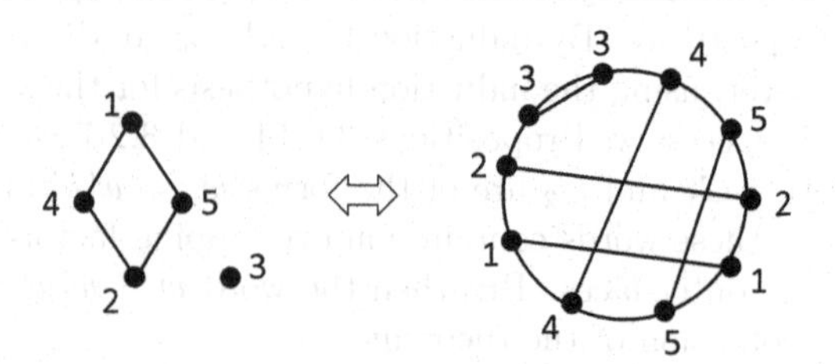

Figure 5.2: A circle graph on five vertices

Corollary 5.1.8. *Each outerplanar graph is a circle graph.*

5.2 On Graphs with Representation Number 3

Unlike the cases of representation numbers 1 and 2, we do not have a characterization of graphs with representation number 3. However, we have a number of results on this class of graphs, which we discuss in this section.

5.2.1 The Petersen Graph

It has already been mentioned above that the Petersen graph is 3-word-representable. The following theorem shows that it is not 2-word-representable, and thus it belongs to $\mathcal{R}_3$.

Theorem 5.2.1. *([73], Konovalov, Linton) The Petersen graph's representation number is* 3.

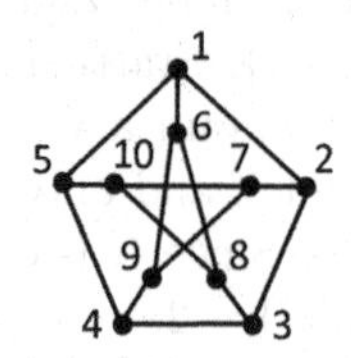

Figure 5.3: The Petersen graph

Proof. In our proof we refer to the Petersen graph in Figure 5.3 (also presented in Figure 3.1). In 2010, Alexander Konovalov and Steve Linton, using the constraint solver Minion, found two 3-uniform words representing the Petersen graph, which are straightforward to check for the right alternation properties:

- 1387296(10)7493541283(10)7685(10)194562 and
- 134(10)58679(10)273412835(10)6819726495.

Konovalov and Linton also checked, using their software, that the graph is not 2-word-representable. Here, we provide a proof of this fact that originally appeared in [73].

Suppose that the graph is 2-word-representable and w is a 2-uniform word representing it. Let x be a letter in w such that there is a minimal number of letters between the two occurrences of x. Since Petersen's graph is regular of degree 3, it is not difficult to see that there must be exactly three letters, which are all different, between the xs (having more letters between xs would lead to having two equal letters there, contradicting the choice of x).

By symmetry, we can assume that $x = 1$, and by Proposition 3.2.7 we can assume that w starts with 1. So, the letters 2, 5 and 6 are between the two 1s, and because of symmetry, the fact that Petersen's graph is *edge-transitive* (that is, each of its edges can be made "internal"), and taking into account that the vertices 2, 5 and 6 are pairwise non-adjacent, we can assume that $w = 12561w_1 6w_2 5w_3 2w_4$ where the w_is are some, possibly empty words for $i \in \{1, 2, 3, 4\}$. To alternate with 6 but not to alternate with 5, the letter 8 must occur in w_1 and w_2. Also, to alternate with 2 but not to alternate with 5, the letter 3 must occur in w_3 and w_4. But then 8833 is a subsequence in w, and thus 8 and 3 must be non-adjacent in the graph, a contradiction. □

5.2.2 Subdivisions of Graphs

The following theorem gives a useful tool for constructing 3-word-representable graphs, that is, graphs with representation number at most 3.

Theorem 5.2.2. *([92]) Let $G = (V, E)$ be a 3-word-representable graph and $x, y \in V$. Denote by H the graph obtained from G by adding to it a path of length at least 3 connecting x and y. Then H is also 3-word-representable.*

Proof. By Theorem 5.4.1 to be discussed in Subsection 5.4.3, we can always attach a leaf (a vertex of degree 1) to any vertex of a 3-word-representable graph obtaining as the result a 3-word-representable graph. Therefore, it is enough to prove the theorem for a path of length 3. In this case, assuming that the path contains two new vertices u and v, $V(H) = V(G) \cup \{u, v\}$ and $E(H) = E(G) \cup \{xu, uv, vy\}$. Let w be a 3-uniform word representing G. Without loss of generality, we may assume that $x^1 < y^1$, where x^i (resp., y^i) denotes the ith occurrence (from left to right) of the letter x (resp., y) in w, and $a < b$ means that the letter a occurs to the left of the letter b in w. There are ten permutations of $\{x^1, x^2, x^3, y^1, y^2, y^3\}$ satisfying the constraints, which we subdivide into five different cases. We consider only one of these cases in detail, since the other cases can be analyzed similarly.

1. $y^2 < x^2$ and $y^3 < x^3$. These conditions cover the permutations

$$x^1y^1y^2y^3x^2x^3 \text{ and } x^1y^1y^2x^2y^3x^3.$$

We substitute y^2 by $v^1u^1y^2v^2$ and x^3 by $u^2x^3v^3u^3$ in w to obtain a 3-uniform word w' representing H. Indeed, since $w' \setminus \{u,v\} = w$, we need only to check the neighbourhoods of u and v in H. These vertices are adjacent in H because they alternate in w'. By Proposition 3.0.15, additionally u could only be adjacent to x, while v could only be adjacent to y. The edges ux and vy indeed exist in H because of the corresponding subsequences $x^1u^1x^2u^2x^3u^3$ and $y^1v^1y^2v^2y^3v^3$ in w'.

2. $y^2 < x^2$ but $x^3 < y^3$. These conditions cover the permutation

$$x^1y^1y^2x^2x^3y^3.$$

We substitute y^1 by $v^1u^1y^1v^2$ and x^3 by $u^2x^3v^3u^3$ in w to obtain the word w' 3-representing H.

3. $x^2 < y^2 < x^3$. These conditions cover the permutations

$$x^1y^1x^2y^2x^3y^3,\ x^1y^1x^2y^2y^3x^3,\ x^1x^2y^1y^2x^3y^3 \text{ and } x^1x^2y^1y^2y^3x^3.$$

We substitute x^1 by $u^1v^1x^1u^2$ and y^2 by $v^2y^2u^3v^3$ in w to obtain the word w' 3-representing H.

4. $x^3 < y^2$ and $x^2 < y^1$. These conditions cover the permutations

$$x^1x^2y^1x^3y^2y^3 \text{ and } x^1x^2x^3y^1y^2y^3.$$

We substitute x^2 by $u^1x^2v^1u^2$ and y^2 by $v^2u^3y^2v^3$ in w to obtain the word w' 3-representing H.

5. $x^3 < y^2$ but $y^1 < x^2$. These conditions cover the permutation

$$x^1y^1x^2x^3y^2y^3.$$

We substitute x^2 by $u^1x^2v^1u^2$ and y^3 by $v^2u^3y^3v^3$ in w to obtain the word w' 3-representing H.

We are done. □

Remark 5.2.3. The third graph in Figure 3.6 shows that it is not possible to reduce the length of the path from 3 to 2 in Theorem 5.2.2. Indeed, removing the vertices 2, 3 and 4 from that graph, we obtain the complete graph K_4, which can be 3-represented, e.g. by the word 156715671567. If the statement of Theorem 5.2.2 were true for paths of length 2, we could insert back, one by one, the removed vertices (together with the respective edges) and obtain a 3-word-representable graph contradicting the fact that the graph in question is non-word-representable.

Definition 5.2.4. A *subdivision* of a graph G is a graph obtained from G by replacing each edge xy in G by a *simple* path (that is, a path without self-intersection) from x to y. A subdivision is called a *k-subdivision* if each of these paths is of length at least k.

Remark 5.2.5. Instead of replacing edges by simple paths while talking about subdivisions, we can think of adding new vertices to the graph, placing them on edges.

Example 5.2.6. A 3-subdivision of the wheel W_5 is presented in Figure 5.4.

Figure 5.4: A 3-subdivision of the wheel W_5

Definition 5.2.7. An *edge contraction* is an operation which removes an edge from a graph while gluing the two vertices it used to connect. An undirected graph G is a *minor* of another undirected graph H if a graph isomorphic to G can be obtained from H by contracting some edges, deleting some edges, and deleting some isolated vertices.

The following theorem follows directly from Theorem 5.2.2 and Definition 5.2.4.

Theorem 5.2.8. *([92]) For every graph G there exists a 3-word-representable graph H that contains G as a minor. In particular, a 3-subdivision of every graph G is 3-word-representable.*

Proof. Suppose that $V(G) = \{1, \ldots, n\}$. Remove all edges from G to obtain the independent set on n vertices, which can be 3-represented by the word $w = 12 \cdots n1122 \cdots nn$. Now insert simple paths of length at least 3, one by one, between vertices x and y whenever $xy \in E(G)$ while extending the word w in a 3-uniform way using the methods in the proof of Theorem 5.2.2. □

Remark 5.2.9. It is straightforward from Theorem 5.2.2 and our proof of Theorem 5.2.8 that a graph obtained from an edgeless graph by inserting simple paths of length at least 3 is 3-word-representable.

Remark 5.2.10. Suppose that each occurrence of "3-word-representable" in the statement of Theorem 5.2.8 is replaced by "word-representable". Clearly, the modified theorem still holds. There is a simple proof of this theorem in terms of semi-transitive orientations. Indeed, make G's original vertices be sources (all the edges are oriented to come out of them) and orient any other edge in an arbitrary way. Then, clearly, no directed path passes through two of G's original vertices, and there is no directed path from a new (just added) vertex to an original vertex of G. These observations confirm that the obtained orientation is semi-transitive, and thus, the obtained graph is word-representable by Theorem 4.1.8.

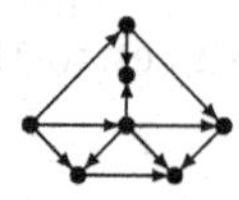
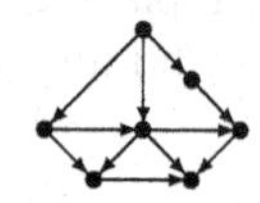

Figure 5.5: Semi-transitive orientations of two graphs from which W_5 is obtained by an edge contraction

Proposition 5.2.11. *If W_5 is obtained from a graph G by an edge contraction then G is word-representable.*

Proof. Without loss of generality, G is one of the two graphs, if orientations are ignored, in Figure 5.5. It is straightforward to check that the orientations in Figure 5.5 are semi-transitive, and thus in either case G is word-representable by Theorem 4.1.8. □

5.2.3 Prisms and Ladder Graphs

Definition 5.2.12. A *prism* Pr_n is a graph consisting of two cycles $12\cdots n$ and $1'2'\cdots n'$, where $n \geq 3$, connected by the edges ii' for $i = 1, \ldots, n$. In particular, the three-dimensional cube is a prism.

Example 5.2.13. Examples of prisms are given in Figure 5.6. The leftmost prism there is called the *triangular prism.* The middle prism is the 3-dimensional cube.

Definition 5.2.14. The *ladder graph* L_n with $2n$ vertices and $3n-2$ edges is presented in Figure 5.7.

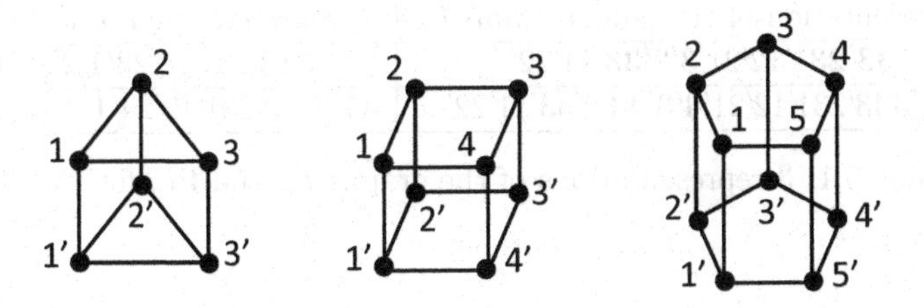

Figure 5.6: Examples of prisms

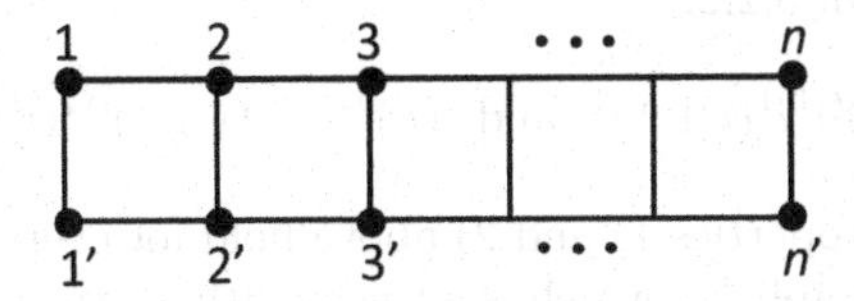

Figure 5.7: The ladder graph L_n

Theorem 5.2.15. *([92]) Every prism Pr_n is 3-word-representable.*

Proof. Our strategy is first to prove that, for $n \geq 2$, the ladder graph L_n can be 3-represented by the word w_n to be constructed below, and then to show how to modify this word to obtain a 3-representation of the prism Pr_n.

We begin with the 3-uniform word

$$w_2 = 121'12'21'2'11'22'$$

representing the 4-cycle $122'1'$. Note that w_2 contains the factors $1^1 2^1$ and $(1')^2(2')^2$, where a^i denotes the ith occurrence of a letter a in the word in question. Add the path $233'2'$ to the 4-cycle using the third case in the proof of Theorem 5.2.2, namely the substitutions

$$2^1 \to 3^1(3')^1 2^1 3^2 \quad \text{and} \quad (2')^2 \to (3')^2(2')^2 3^3 (3')^3$$

in w_2 to obtain the 3-uniform word

$$w_3 = 133'231'12'21'3'2'33'11'22',$$

which satisfies the following properties for $i = 3$:

1) w_i contains the factors $1^1 i^1$ and $(1')^2(i')^2$;

2) The subword of w_i induced by the letters 1 and i is $1^1 i^1 i^2 1^2 i^3 1^3$, while the subword induced by $1'$ and i' is $(i')^1(1')^1(1')^2(i')^2(i')^3(1')^3$.

n	3-representation of the ladder graph L_n	3-representation of the prism Pr_n
3	$1\mathbf{33'23}1'12'21'\mathbf{3'2'33'}11'22'$	$\mathbf{31}3'231'12'2\mathbf{3'1'}2'33'11'22'$
4	$1\mathbf{44'34}3'231'12'21'\mathbf{4'3'44'}2'33'11'22'$	$\mathbf{41}4'343'231'12'2\mathbf{4'1'}3'44'2'33'11'22'$

Table 5.1: 3-representations of the graphs L_n and Pr_n for $n = 3, 4$

Repeat the operation of adding a path $i(i+1)(i+1)'i'$ for $i = 3, 4, \ldots, n-1$. Since $(i, i') \in E(\mathrm{Pr}_n)$ we can always do this using the substitution in the third case in the proof of Theorem 5.2.2:

$$i^1 \to (i+1)^1((i+1)')^1 i^1 (i+1)^2 \quad \text{and} \quad (i')^2 \to ((i+1)')^2(i')^2(i+1)^3((i+1)')^3.$$

It is easy to see that properties 1) and 2) above hold for every i. Thus, the word w_n represents the ladder graph L_n, which is a prism without the edges $1n$ and $1'n'$. Now substitute the factors 1^1n^1 and $(1')^2(n')^2$ in w_n by $n^1 1^1$ and $(n')^2(1')^2$, respectively. The obtained word represents the prism Pr_n. Indeed, due to property 2), $(1, n)$ and $(1', n')$ become edges, and all the other adjacencies in the graph L_n are not changed, since the subwords induced by any other pair of letters remain the same. □

Example 5.2.16. In Table 5.1, we present 3-representations of the graphs L_n and Pr_n for $n = 3, 4$. The representations are obtained following the steps described in the proof of Theorem 5.2.15. The respective substitutions are marked in bold.

The following statement is an immediate corollary to Theorem 5.2.15.

Corollary 5.2.17. *We have $\mathcal{R}(Pr_n) \leq 3$, that is, each prism's representation number is at most* 3.

In the following two theorems we prove that each prism's representation number is exactly 3.

Theorem 5.2.18. *([92]) The representation number of the triangular prism Pr_3 is* 3.

Proof. By Theorem 5.2.15, it remains to show that Pr_3 is not 2-word-representable. In our proof we refer to the leftmost graph in Figure 5.6.

Suppose that Pr_3 is 2-word-representable by a word w. Denote by x a letter in w such that no letter occurs twice between x^1 and x^2, the first and second occurrences of x in w, respectively. It is not difficult to see that such a letter x must exist. By Proposition 3.0.15 and the definition of Pr_n, there are exactly three neighbours of x between x^1 and x^2. Due to the symmetry and Proposition 3.2.7, we may assume that the word starts with x and $x = 1$. By Proposition 3.0.14 and the symmetry of the vertices 2 and 3, there are two non-equivalent cases, where by $a < b$ we mean that the letter a is to the left of letter b in w:

- The word w starts with $121'31 = 1^1 2^1 (1')^1 3^1 1^2$. Then $(1')^2 < 2^2$, $(1')^2 > 3^2$ and $2^2 < 3^2$ since, respectively, $21' \notin E(\mathrm{Pr}_3)$, $31' \notin E(\mathrm{Pr}_3)$ and $23 \in E(\mathrm{Pr}_3)$, which is impossible.
- The word w starts with $1231'1 = 1^1 2^1 3^1 (1')^1 1^2$. Then $(1')^2 < 2^2 < 3^2$, because $21' \notin E(\mathrm{Pr}_3)$ and $23 \in E(\mathrm{Pr}_3)$. Since $3'$ is adjacent to both $1'$ and 3, we have $(3')^1 < (1')^2$ and $(3')^2 > 3^2$. But then 2 and $3'$ alternate, a contradiction to $23' \notin E(\mathrm{Pr}_3)$. □

Theorem 5.2.19. *([89]) For $n \geq 4$, $\mathcal{R}(Pr_n) = 3$.*

Proof. By Theorem 5.2.15, it remains to show that for $n \geq 4$, Pr_n is not 2-word-representable.

Suppose that Pr_n can be 2-represented by a word w for $n \geq 4$. It is not difficult to see that there must exist a letter x in w such that no letter occurs twice between the two copies of x. By Proposition 3.0.15, there are exactly three letters between the copies of x. Using symmetry, Proposition 3.2.7 and Proposition 3.0.14, we only need to consider two cases (the second one is unnecessary in the case of $n = 4$ because of symmetry) where we take into account that the vertices $1'$, 2 and n form an independent set:

- w is of the form $11'2n1 \cdots n \cdots 2 \cdots 1' \cdots$. Since $nn' \in E(\mathrm{Pr}_n)$ and $1'n' \in E(\mathrm{Pr}_n)$, we can refine the structure of w as follows

$$w = 11'2n1 \cdots n' \cdots n \cdots 2 \cdots 1' \cdots n' \cdots .$$

  However, 2 and n' alternate in w contradicting the fact that $2n' \notin E(\mathrm{Pr}_n)$.
- w is of the form $121'n1 \cdots n \cdots 1' \cdots 2 \cdots$. In this case, we will refine the structure of w in two different ways and then merge these refinements:
  - Since $22' \in E(\mathrm{Pr}_n)$, $1'2' \in E(\mathrm{Pr}_n)$ and $2'n \notin E(\mathrm{Pr}_n)$, w must be of the form

$$w = 121'n1 \cdots n \cdots 2' \cdots 1' \cdots 2 \cdots 2' \cdots .$$

  - Since $nn' \in E(\mathrm{Pr}_n)$, $1'n' \in E(\mathrm{Pr}_n)$ and $2n' \notin E(\mathrm{Pr}_n)$, w must be of the form

$$w = 121'n1 \cdots n' \cdots n \cdots 1' \cdots n' \cdots 2 \cdots$$

Merging the refinements, we see that w must be of the form

$$w = 121'n1 \cdots n' \cdots n \cdots 2' \cdots 1' \cdots n' \cdots 2 \cdots 2' \cdots .$$

However, we see that the letters $2'$ and n' alternate in w contradicting the fact that $2'n' \notin E(\mathrm{Pr}_n)$. □

Remark 5.2.20. ([89]) Applying Proposition 3.3.6 as many times as necessary, we see that if $G \in \mathcal{R}_3$ then a graph obtained from G by attaching simple paths of any lengths to vertices in G belongs to $\mathcal{R}_3$; call this way to extend graphs "operation 1". Also, by Theorem 5.2.2, we can add simple paths of length at least 3 connecting any pair of vertices in $G \in \mathcal{R}_3$ and still obtain a graph in $\mathcal{R}_3$ (if a 2-word-representable graph were to be obtained, we would have a contradiction with $G \in \mathcal{R}_3$); call this way to extend graphs "operation 2". The only members of $\mathcal{R}_3$ known to us are the Petersen graph, prisms and any other graph obtained from these by applying operations 1 and 2 an arbitrary number of times in any order.

We end this subsection by considering ladder graphs.

It is clear that $L_1 = K_2$ and thus $\mathcal{R}(L_1) = 1$ (see Theorem 5.1.1), while $L_2 = C_4$ and thus $\mathcal{R}(L_2) = 2$ (see the way to represent cycles discussed in Subsection 3.2 or Theorem 5.1.7). Moreover, from our proof of Theorem 5.2.15 it follows that the ladder graph L_n is 3-word-representable for $n \geq 3$, and thus $\mathcal{R}(L_n) \leq 3$. The following theorem shows that for $n \geq 2$, $\mathcal{R}(L_n) = 2$.

Theorem 5.2.21. *([89]) For $n \geq 2$, $\mathcal{R}(L_n) = 2$.*

Proof. Recall that the graph L_n is presented in Figure 5.7. We prove the statement by induction on n.

L_1 can be 2-represented by the word $w_1 = 11'11'$, which has the factor $1'1$. Substituting the factor $1'1$ in w_1 by $2'1'22'12$ and reversing the entire word, we obtain the word $w_2 = 1'212'21'2'1$, which contains the factor $2'2$. It is straightforward to check that w_2 represents L_2 since one only needs to check the alternation properties of the just added letters 2 and $2'$.

More generally, given a 2-representation w_i of L_i containing the factor $i'i$, we substitute $i'i$ in w_i by $(i+1)'i'(i+1)(i+1)'i(i+1)$ and reverse the entire word to obtain the word w_{i+1} containing the factor $(i+1)'(i+1)$. It is straightforward to check that w_{i+1} represents L_{i+1} since the only thing that needs to be checked is the right alternation properties of the just added letters $i+1$ and $(i+1)'$. We are done. □

Example 5.2.22. In Table 5.2, we present 2-representations of the ladder graph L_n for $n = 1, \ldots, 5$. The factors $n'n$ are indicated in bold.

n	2-representation of the ladder graph L_n
1	$11'11'$
2	$1'212'21'2'1$
3	$12'1'323'32'3'121'$
4	$1'213'2'434'43'4'231'2'1$
5	$12'1'324'3'545'54'5'342'3'121'$

Table 5.2: 2-representations of the ladder graph L_n for $n = 1, \ldots, 5$

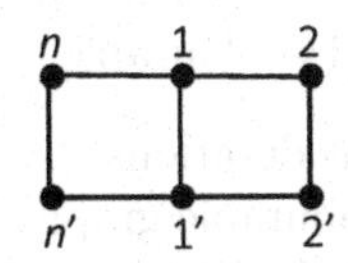

Figure 5.8: The ladder graph on 6 vertices

Remark 5.2.23. By Theorems 5.1.7 and 5.2.21, any ladder graph is a circle graph.

Remark 5.2.24. ([89]) The thoughtful reader will notice what seems to be an inconsistency: our argument in the proof of Theorem 5.2.19 seems to show that the graph in Figure 5.8 is not 2-word-representable, while by Theorem 5.2.21 we see that this graph is 2-word-representable by $n1'n'212'21'2'n1n'$ (we renamed the labels in L_3 and used the third line in Table 5.2). The reason for the possible confusion is that we used symmetry in the proof of Theorem 5.2.19 to assume that there are exactly three letters between the 1s; such an assumption cannot be made when dealing with the graph in Figure 5.8 because 1 is an "internal" vertex there, while there are "external" vertices as well, namely $2, 2', n, n'$.

5.2.4 Chromatic Number of Graphs in $\mathcal{R}_3$

Theorems 5.1.1 and 5.1.7 show that the chromatic number of 1- or 2-word-representable graphs is not bounded by any constant. Indeed, K_n is n-colourable, while circle graphs formed by K_{n-1} and an isolated vertex (which are not 1-word-representable) are $(n-1)$-colourable. The following theorem shows that this is also the case for graphs in $\mathcal{R}_3$.

Theorem 5.2.25. *([89]) Graphs in $\mathcal{R}_3$ can have arbitrarily large chromatic number.*

Proof. Suppose that any graph in class $\mathcal{R}_3$ is c-colourable for some constant c. Consider the triangular prism Pr_3 to the left in Figure 5.14, which belongs to $\mathcal{R}_3$ by Theorem 5.2.18. Replace the vertex 1 in Pr_3 with a module K_{c+1}, the complete graph on $c+1$ vertices (see Subsection 5.4.4 for the notion of a module) as shown for the case of $c = 2$ in Figure 5.14. Denote the obtained graph by Pr'_3. Since $\mathcal{R}(K_{c+1}) = 1$, by Theorem 5.4.7, $\mathrm{Pr}'_3 \in \mathcal{R}_3$. However, Pr'_3 is not c-colourable since it contains a clique of size $c+1$ (Pr'_3 is $(c+1)$-colourable). We obtain a contradiction with our assumption. □

Remark 5.2.26. ([89]) We note that an alternative way to obtain a contradiction in the proof of Theorem 5.2.25 is to consider the triangular prism Pr_3 (which is in $\mathcal{R}_3$), the complete graph K_{c+1} (which is in $\mathcal{R}_1$), and either to connect these graphs by an edge, or glue these graphs at a vertex. Then by Theorem 5.4.3, the obtained graph will be in $\mathcal{R}_3$, but it is $(c+1)$-colourable.

Remark 5.2.27. ([89]) From considerations in this subsection it follows that, in particular, the classes $\mathcal{R}_3$ and 3-colourable graphs (which are word-representable by Theorem 4.3.3) are not comparable (neither of these classes is included in the other one).

5.3 Asymptotic Enumeration

We now apply the results of Section 2.3 in order to derive an asymptotic formula for the number of word-representable graphs. We refer the reader to that section for notations and definitions used in the proof. We start with the number of n-vertex *labelled* word-representable graphs, which we denote by b_n.

Theorem 5.3.1.

$$\lim_{n\to\infty} \frac{\log_2 b_n}{\binom{n}{2}} = \frac{2}{3}.$$

Proof. By Theorem 4.3.3, $\mathcal{E}_{3,0}$ is a subclass of the class of word-representable graphs and hence its index is at least 3. In order to show that the index does not exceed 3, we observe that the graph A represented in Figure 4.4 belongs to all minimal classes of index 4, and hence the family of word-representable graphs does not contain any of these minimal classes by Theorem 4.1.8 and the fact that A is not semi-transitively orientable (see the proof of Theorem 4.5.1). Therefore, the index of the class of word-representable graphs is precisely 3. □

We now proceed to the number of unlabelled n-vertex word-representable graphs, which we denote by a_n.

Theorem 5.3.2.

$$\lim_{n\to\infty} \frac{\log_2 a_n}{\binom{n}{2}} = \frac{2}{3}.$$

Proof. Clearly, $b_n \leq n!a_n$ and $\log_2 n! \leq \log_2 n^n = n \log_2 n$. Therefore,

$$\lim_{n\to\infty} \frac{\log_2 b_n}{\binom{n}{2}} \leq \lim_{n\to\infty} \frac{\log_2(n!a_n)}{\binom{n}{2}} = \lim_{n\to\infty} \frac{\log_2 n! + \log_2 a_n}{\binom{n}{2}} \leq \lim_{n\to\infty} \frac{n\log_2 n + \log_2 a_n}{\binom{n}{2}}$$
$$= \lim_{n\to\infty} \frac{\log_2 a_n}{\binom{n}{2}}.$$

On the other hand, obviously $b_n \geq a_n$ and hence $\lim_{n\to\infty} \frac{\log_2 b_n}{\binom{n}{2}} \geq \lim_{n\to\infty} \frac{\log_2 a_n}{\binom{n}{2}}$. Combining, we obtain $\lim_{n\to\infty} \frac{\log_2 b_n}{\binom{n}{2}} = \lim_{n\to\infty} \frac{\log_2 a_n}{\binom{n}{2}}$. Together with Theorem 5.3.1, this proves the result. □

Corollary 5.3.3.

$$a_n = 2^{\frac{n^2}{3}+o(n^2)}.$$

5.4 On Graph Operations Preserving (Non-)word-representability

In this section we consider some of the most basic operations on graphs and discuss what can be said about these operations with respect to preserving (non-)word-representability. The operations considered by us are taking the complement, edge contraction, connecting two graphs by an edge and gluing two graphs in a clique, replacing a vertex with a module, Cartesian product, rooted product and taking line graph.

5.4.1 Taking the Complement

Starting with a word-representable graph and taking its complement, we may either obtain a word-representable graph or not. Indeed, for example, the complement of the complete graph K_3 on vertices 1, 2 and 3, which can be represented by 123, is an edgeless graph, which can be represented by 112233. On the other hand, let G be the graph formed by the 5-cycle (2,4,6,3,5) and an isolated vertex 1. Using the technique to represent cycles presented in Section 3.2, the 5-cycle can be represented by the word 2542643653 and thus the graph G can be represented by the word 112542643653. However, taking the complement of G_1, we obtain the wheel W_5 presented in Figure 3.6, which is not word-representable.

Similarly, starting with a non-word-representable graph and taking its complement, we can either obtain a word-representable graph or not. Indeed, the complement of the non-word-representable wheel W_5 is word-representable, as is discussed above. On the other hand, the graph G having two connected components, one W_5 and the other one the 5-cycle C_5, is non-word-representable because of the induced subgraph W_5, while the complement of G also contains an induced subgraph W_5 (formed by the vertices of C_5 in G and any of the remaining vertices) and thus is also non-word-representable.

5.4.2 Edge Contraction

Recall the definition of edge contraction in Definition 5.2.7. Edge contraction does not preserve the property of being a word-representable graph as is shown by Proposition 5.2.11, although in many cases, for example in the case of paths, we do obtain word-representable graphs from word-representable graphs when contracting edges.

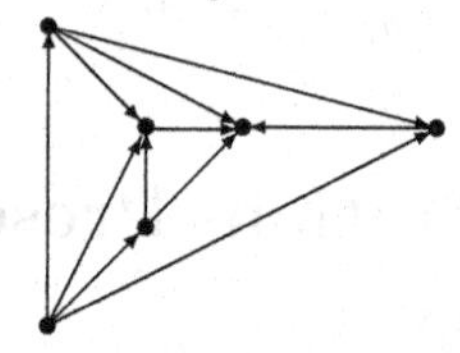

Figure 5.9: A semi-transitive orientation of the graph obtained from the graph in Figure 4.4 by edge contraction of the edge 25

When starting from a non-word-representable graph, a graph obtained from it by edge contraction can be either word-representable or non-word-representable. For example, contracting the edge 25 in the non-word-representable graph in Figure 4.4, we obtain a word-representable graph whose semi-transitive orientation is presented in Figure 5.9. On the other hand, attaching a path, say of length 3, to any vertex of the graph in Figure 4.4, and contracting any edge on the path, we obtain a non-word-representable graph from a non-word-representable graph.

5.4.3 Connecting Two Graphs by an Edge and Gluing Two Graphs in a Clique

The operations of connecting two graphs, G_1 and G_2, by an edge and gluing two graphs at a vertex are presented schematically in Figure 5.10. It follows directly from Theorem 4.1.8 that if both G_1 and G_2 are word-representable then the resulting graphs will be word-representable too, while if at least one of G_1 or G_2 is non-word-representable then the resulting graphs will be non-word-representable. Indeed, if G_1 and G_2 are oriented semi-transitively, then orienting the edge xy in either direction will give no chance for the resulting graph to have a shortcut (see Remark 4.1.3

for the notion of a shortcut) thus resulting in a semi-transitively oriented graph; similarly, no shortcut is possible when semi-transitively oriented G_1 and G_2 are glued at a vertex z.

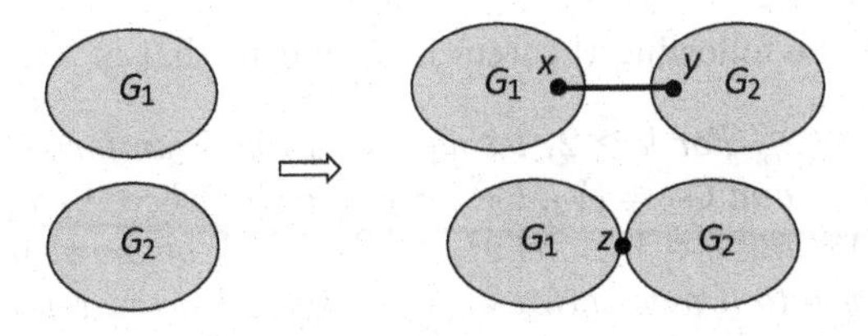

Figure 5.10: Connecting graphs by an edge and gluing graphs at a vertex

While the arguments above involving orientations answer the question on word-representability of connecting graphs by an edge or gluing graphs at a vertex, they do not allow us to answer the following question: if G_1 is k_1-word-representable, G_2 is k_2-word-representable, and G is k-word-representable (such a k must exist by Theorem 3.3.1) then what can be said about k? Theorem 5.4.3 below, which is based on Theorems 5.4.1 and 5.4.2, answers this question.

Recall from Definition 3.0.1 that $A(w)$ is the set of letters occurring in a word w.

Theorem 5.4.1. *([92]) For $k \geq 2$, let w_1 and w_2 be k-uniform words representing graphs $G_1 = (V_1, E_1)$ and $G_2 = (V_2, E_2)$, respectively, where V_1 and V_2 are disjoint. Suppose that $x \in V_1$ and $y \in V_2$. Let H_1 be the graph $(V_1 \cup V_2, E_1 \cup E_2 \cup \{xy\})$. Then H_1 is k-word-representable.*

Proof. By Proposition 3.2.7, we may assume that $w_1 = A_1x^1A_2x^2 \cdots A_kx^k$ and $w_2 = y^1B_1y^2B_2 \cdots y^kB_k$, where $A_1 \cdots A_k = w_1 \setminus \{x\}$ and $B_1 \cdots B_k = w_2 \setminus \{y\}$, and x^i and y^i are the ith occurrence of x and y in w_1 and w_2, respectively. Then the word

$$w_3 = A_1x^1A_2y^1x^2B_1A_3 \cdots y^{k-2}x^{k-1}B_{k-2}A_ky^{k-1}x^kB_{k-1}y^kB_k$$

represents H_1.

Indeed, it is easy to see that x and y alternate in w_3, and thus they are connected by an edge in H_1. Also, note that $w_3 \setminus A(w_2) = w_1$ and $w_3 \setminus A(w_1) = w_2$. Therefore, for $i = 1, 2$, the graph induced by $A(w_i)$ is isomorphic to G_i, and it is enough to show that there are no edges between these graphs except for xy.

Let $a \in V_1, b \in V_2$, and $\{a, b\} \neq \{x, y\}$, say, $a \neq x$. Since w_1 is k-uniform, there are k occurrences of a in w_1. Then, by the pigeonhole principle, among the $k-1$ subsets $A_1 \cup A_2, A_3, A_4, \ldots, A_k$ there exists one containing at least two copies of a. Hence, the word induced by a and b in w_3 contains aa as a factor, and

$ab \notin E(H_1)$. Similarly, if $b \neq y$ (while $a = x$), then among the $k-1$ subsets $B_1, B_2, \ldots, B_{k-2}, B_{k-1} \cup B_k$ there exists one containing at least two copies of b. Hence, the word induced by a and b in w_3 contains bb as a factor, and $ab \notin E(H_1)$. □

Our proof of the following theorem is similar to that of Theorem 5.4.1.

Theorem 5.4.2. *([92]) For $k \geq 2$, let w_1 and w_2 be k-uniform words representing graphs $G_1 = (V_1, E_1)$ and $G_2 = (V_2, E_2)$, respectively, where V_1 and V_2 are disjoint. Suppose that $x \in V_1$ and $y \in V_2$. Let H_2 be the graph obtained from G_1 and G_2 by identifying x and y into a new vertex z. Then H_2 is k-word-representable.*

Proof. By Proposition 3.2.7, we may assume that $w_1 = A_1x^1A_2x^2 \cdots A_kx^k$ and $w_2 = y^1B_1y^2B_2 \cdots y^kB_k$, where $A_1 \cdots A_k = w_1 \setminus \{x\}$ and $B_1 \cdots B_k = w_2 \setminus \{y\}$, and x^i and y^i are the ith occurrence of x and y in w_1 and w_2, respectively. Then the word

$$w_4 = A_1z^1A_2B_1z^2A_3B_2z^3 \cdots A_kB_{k-1}z^kB_k$$

represents H_2 (where z^i is the ith occurrence of the letter z in w_4).

Indeed, it is not difficult to see that z alternates with a letter s in w_4 if and only if x alternates with s in w_1 or y alternates with s in w_2. Thus, any edge with an endpoint z in H_2 is necessarily an edge with an endpoint x in G_1 or an edge with an endpoint y in G_2.

On the other hand, if $a \in V_1$, $a \neq x$, and $b \in V_2$, $b \neq y$, then $ab \notin E(H_2)$. Indeed, since w_1 is k-uniform, there are k occurrences of a in w_1. Then, by the pigeonhole principle, among the $k-1$ subsets $A_1 \cup A_2, A_3, A_4, \ldots, A_k$ there exists one containing at least two copies of a. Hence, the word induced by a and b in w_4 contains aa as a factor, and $ab \notin E(H_2)$. □

Theorem 5.4.3. *([89]) Suppose that for graphs $G_1 = (V_1, E_1)$ and $G_2 = (V_2, E_2)$, $\mathcal{R}(G_1) = k_1$ and $\mathcal{R}(G_2) = k_2$, $x \in V_1$, $y \in V_2$ and $k = \max(k_1, k_2)$. Also, let the graph G' be obtained by connecting G_1 and G_2 by the edge xy, and the graph G'' be obtained from G_1 and G_2 by identifying the vertices x and y into a single vertex z. The following holds.*

1. *If $|V_1| = |V_2| = 1$ then both G' and G'' are cliques and thus $k_1 = k_2 = 1$. In this case, $\mathcal{R}(G') = \mathcal{R}(G'') = 1$.*

2. *If $\min(|V_1|, |V_2|) = 1$ but $\max(|V_1|, |V_2|) > 1$ then $\mathcal{R}(G'') = k$ and $\mathcal{R}(G') = \max(k, 2)$.*

3. *If $\min(|V_1|, |V_2|) > 1$ then $\mathcal{R}(G') = \mathcal{R}(G'') = \max(k, 2)$.*

Proof. The first part of the statement is easy to see since both a single vertex (G'') and the one-edge graph (G') are 1-word-representable by Theorem 5.1.1.

For part 2, without loss of generality, $|V_1| = 1$ (that is, $V_1 = \{x\}$) and thus $G'' = G_2$ leading to $\mathcal{R}(G'') = k$. On the other hand, G' is not a clique and thus $\mathcal{R}(G') \geq 2$. If $k_2 = 1$ then G_2 can be represented by the permutation $yy_1 \cdots y_{|V_2|-1}$ for $y, y_i \in V_2$ and thus G' can be represented by $xyxy_1 \cdots y_{|V_2|-1}yy_1 \cdots y_{|V_2|-1}$ leading to $\mathcal{R}(G') = 2$. However, if $k_2 \geq 2$, so that $k = k_2$, we can take any k-word-representation of G_2 and replace in it every other occurrence of the letter y by xyx to obtain a k-word-representation of G'. Thus, $\mathcal{R}(G') = k$ because if it were less than k, we would have $\mathcal{R}(G_2) < k$ by Remark 3.3.5, a contradiction.

For part 3, neither G' nor G'' is a clique and thus $\mathcal{R}(G'), \mathcal{R}(G'') \geq 2$. By Proposition 3.2.13, both G_1 and G_2 are k-word-representable. If $k \geq 2$ then by Theorems 5.4.1 and 5.4.2 both G' and G'' are k-word-representable, leading to $\mathcal{R}(G') = \mathcal{R}(G'') = k$ since if $\mathcal{R}(G') < k$ or $\mathcal{R}(G'') < k$ we would obtain a contradiction either with $\mathcal{R}(G_1) = k_1$ or with $\mathcal{R}(G_2) = k_2$ by Remark 3.3.5. Finally, if $k = 1$ then G_1 and G_2 must be cliques, which can be represented by permutations $x_1 \cdots x_{|V_1|-1}x$ and $yy_1 \cdots y_{|V_2|-1}$, respectively, for $x, x_i \in V_1$ and $y, y_i \in V_2$. Then the words

$$x_1 \cdots x_{|V_1|-1}xx_1 \cdots x_{|V_1|-1}yxy_1 \cdots y_{|V_2|-1}yy_1 \cdots y_{|V_2|-1}$$

and

$$x_1 \cdots x_{|V_1|-1}zx_1 \cdots x_{|V_1|-1}y_1 \cdots y_{|V_2|-1}zy_1 \cdots y_{|V_2|-1}$$

2-word-represent the graphs G' and G'', respectively, and thus $\mathcal{R}(G') = \mathcal{R}(G'') = 2$. □

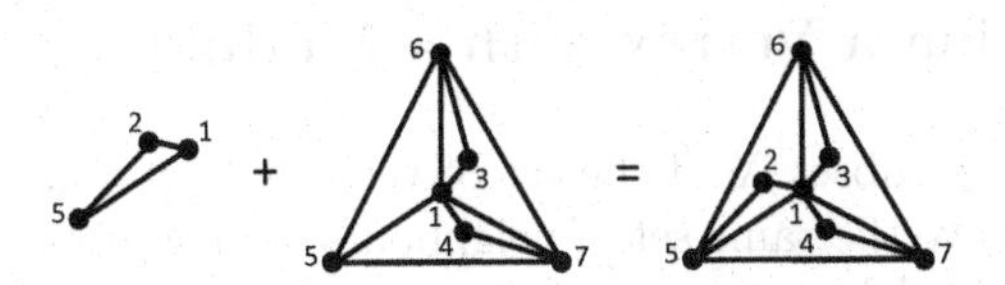

Figure 5.11: Identifying an edge in two graphs

Remark 5.4.4. A vertex is a clique of size 1, and Theorem 5.4.3 shows that identifying cliques of size 1 from two word-representable graphs gives a word-representable graph. This is not necessarily the case with cliques of size more than 1, as remarked in [73]. Indeed, one can see that the third non-word-representable graph in Figure 3.6 is obtained by identifying the edge 15 in the triangle on vertices 1, 2, 5 and the respective edge in another graph as shown in Figure 5.11. The triangle is word-representable, while the other graph used for gluing is word-representable too (which follows from the fact that the wheel W_5 is the only non-word-representable graph on six vertices; alternatively, see the left graph in Figure 5.12 for the respective semi-transitive orientation). For another example, first presented in this book, see Figure 5.13 showing identification of the triangle 145 in two word-representable graphs. The graph on six vertices in Figure 5.13 is word-representable because it is different from W_5, while the resulting graph on seven vertices is non-word-representable as presented in Figure 3.9.

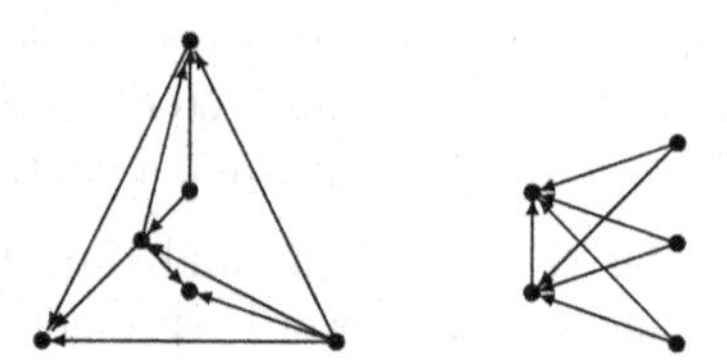

Figure 5.12: Two semi-transitively oriented graphs

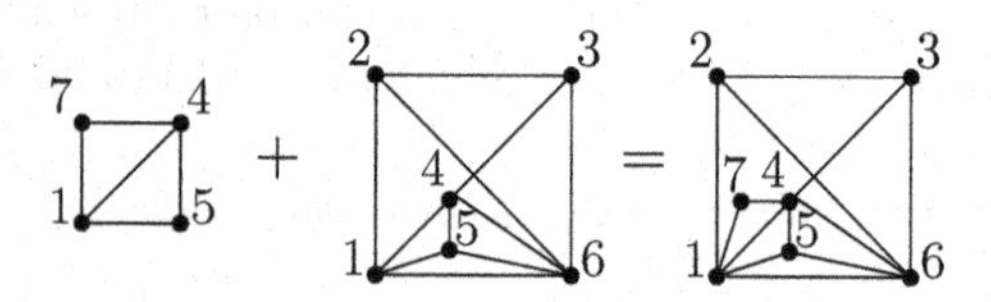

Figure 5.13: Identifying a triangle in two graphs

5.4.4 Replacing a Vertex with a Module

Definition 5.4.5. A subset X of the set of vertices V of a graph G is a *module* if all members of X have the same set of neighbours among vertices not in X (that is, among vertices in $V \setminus X$).

Example 5.4.6. Figure 5.14 shows replacement of the vertex 1 in the triangular prism by the module K_3 formed by the vertices a, b and c.

The following theorem is an extended version of an observation made in [73].

Theorem 5.4.7. *([89]) Suppose that G is a word-representable graph and $x \in V(G)$. Let G' be obtained from G by replacing x with a module M, where M is any comparability graph (in particular, any clique). Then G' is also word-representable. Moreover, if $\mathcal{R}(G) = k_1$ and $\mathcal{R}(M) = k_2$ then $\mathcal{R}(G') = k$, where $k = \max\{k_1, k_2\}$.*

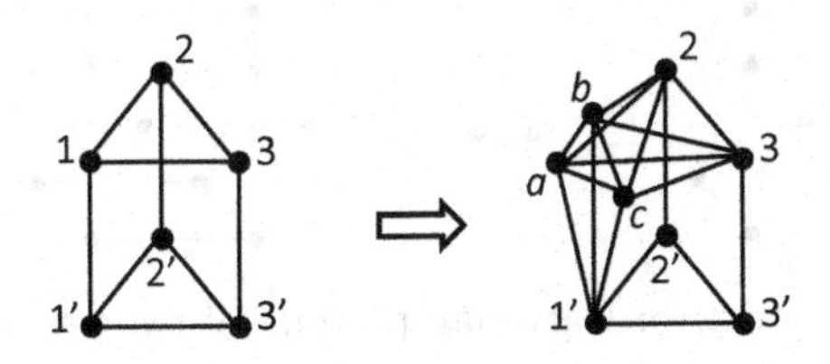

Figure 5.14: Replacing a vertex by a module

Proof. By Theorem 3.4.3, M can be represented by a word $p_1 \cdots p_{k_2}$, where p_i is a permutation (of length equal to the number of vertices in M) for $1 \leq i \leq k_2$. If $k_2 < k$, we can adjoin to the representation of M any number of copies of permutations in the set $\Pi = \{p_1, \ldots, p_{k_2}\}$ to obtain a k-representation $p = p_1 \cdots p_k$ of M, where $p_i \in \Pi$ (indeed, no alternation properties will be changed while adjoining such extra permutations).

Using Proposition 3.2.13, if necessary (when $k_1 < k$), we can assume that G can be k-represented by a word $w = w_0 x w_1 x w_2 \cdots x w_k$ for some words w_i not containing x for all $0 \leq i \leq k$. But then the word $w' = w_0 p_1 w_1 p_2 w_2 \cdots p_k w_k$ k-represents G'. Indeed, it is easy to see that the pairs of letters in w' from $V(M)$ have the right alternation properties, as do the pairs of letters from $V(G) \setminus \{x\}$ in w'. On the other hand, it is not difficult to see that if $y \in V(M)$ and $z \in V(G) \setminus \{x\}$ then $yz \in E(G')$ if and only if $xz \in E(G)$.

If G' were $(k-1)$-word-representable, we would either obtain a contradiction with $\mathcal{R}(G) = k_1$ (after replacing each p_i in w' with x) or with $\mathcal{R}(M) = k_2$ (after removing all letters in w' that are not in M). Thus, $\mathcal{R}(G') = k$. □

5.4.5 Cartesian Product of Two Graphs

Definition 5.4.8. The *Cartesian product* $G \square H$ of graphs $G = (V(G), E(G))$ and $H = (V(H), E(H))$ is a graph such that

- the vertex set of $G \square H$ is the Cartesian product $V(G) \times V(H)$; and
- any two vertices (u, u') and (v, v') are adjacent in $G \square H$ if and only if either
  - $u = v$ and u' is adjacent to v' in H, or
  - $u' = v'$ and u is adjacent to v in G.

Example 5.4.9. See Figure 5.15 for an example of the Cartesian product of two graphs.

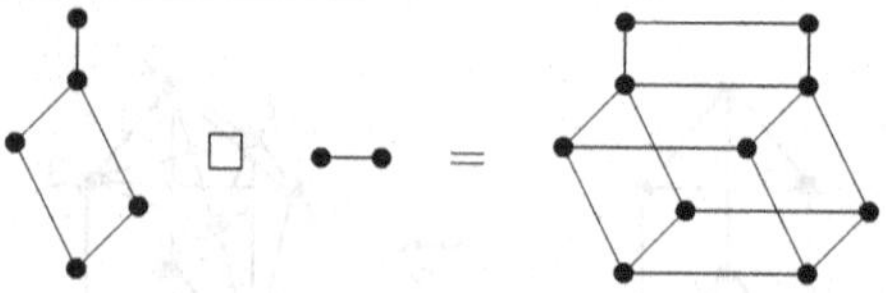

Figure 5.15: Cartesian product of two graphs

A proof of the following theorem was given by Bruce Sagan in 2014. The theorem states that taking the Cartesian product of two word-representable graphs results in a word-representable graph. Note that if one of two graphs is non-word-representable then because of the hereditary property their Cartesian product must be non-word-representable.

Theorem 5.4.10. *(Bruce Sagan) Let G and H be two word-representable graphs. Then the Cartesian product $G \Box H$ is also word-representable.*

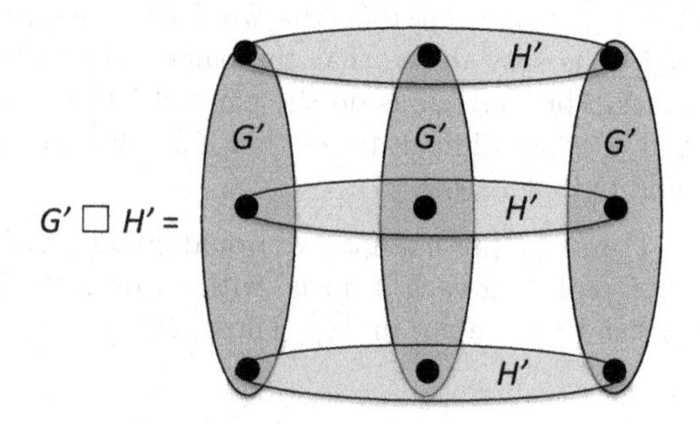

Figure 5.16: Cartesian product of two graphs

Proof. Since G and H are word-representable, we can orient these graphs semi-transitively to obtain directed graphs G' and H', respectively. We claim that the directed graph $G' \Box H'$ presented schematically in Figure 5.16 is oriented semi-transitively, giving us at once that $G \Box H$ is word-representable. We note that all copies of G' (resp., H') must be placed identically to each other in the sense that whenever an edge is oriented upwards/downwards/to the left/to the right in one of the copies of a graph, this edge must be oriented in exactly the same way in any other copy of the graph.

Note that any directed path in $G' \Box H'$ is a sequence of horizontal and vertical steps such that by ignoring all vertical steps one obtains a directed path in H', while by ignoring all horizontal steps one obtains a directed path in G' (this is because we placed copies of the graphs in a uniform way).

If $G'\square H'$ contained a directed cycle, after ignoring all vertical (resp., horizontal) steps we would obtain a directed cycle in H' (resp., G'), which would contradict the fact that G' and H' are semi-transitively oriented. Thus, $G'\square H'$ is acyclic.

If $G'\square H'$ contains a shortcut $v_1v_2\cdots v_k$ with the directed edge from v_1 to v_k, then either the edge v_1v_k is in one of the copies of H', or this edge is in one of the copies of G'. In the first case, ignoring all horizontal steps in $v_1v_2\cdots v_k$ (there must be at least one vertical step since H' is shortcut-free) we obtain a cycle in G'. In the second case, ignoring all vertical steps in $v_1v_2\cdots v_k$ (there must be at least one horizontal step since G' is shortcut-free) we obtain a cycle in H'. In either case, we obtain a contradiction with G' and H' being acyclic. □

5.4.6 Rooted Product of Graphs

Definition 5.4.11. The *rooted product* of a graph G and a rooted graph H, $G \circ H$, is defined as follows: take $|V(G)|$ copies of H, and for every vertex v_i of G, identify v_i with the root vertex of the ith copy of H.

Example 5.4.12. See Figure 5.17 for an example of the rooted product of two graphs.

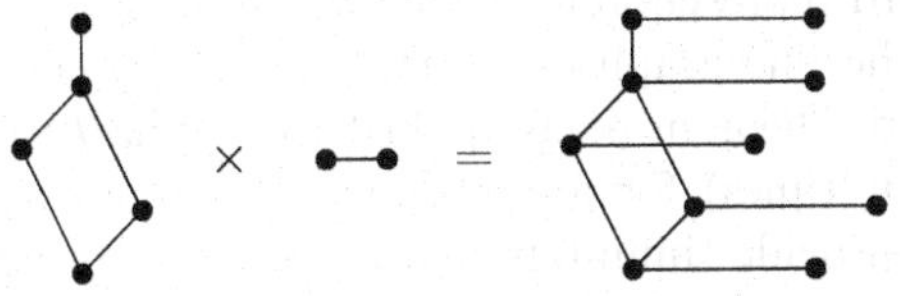

Figure 5.17: Rooted product of two graphs

The next theorem is an analogue of Theorem 5.4.10 for the rooted product of two graphs.

Theorem 5.4.13. *Let G and H be two word-representable graphs. Then the rooted product $G \circ H$ is also word-representable.*

Proof. Identifying a vertex v_i in G with the root vertex of the ith copy of H in the definition of the rooted product gives a word-representable graph by Theorem 5.4.2. Thus, identifying the root vertices, one by one, we will keep obtaining word-representable graphs, which gives us at the end word-representability of $G \circ H$. □

5.4.7 Taking Line Graph Operation

Taking line graph is a graph operation that we will also discuss in Section 5.5. Recall Definition 2.2.27 for the notion of a line graph. While it is easy to give examples of word-representable graphs whose line graphs are also word-representable (e.g. take any graph on at most five edges), it turns out that there are word-representable graphs whose line graphs are not word-representable. An example of such a graph is presented in Figure 5.18. This graph is to the left in that figure, and we label its edges according to its line graph vertices' labels (presented to the right in Figure 5.18). The fact that the original graph is word-representable is demonstrated by its semi-transitive orientation in Figure 5.12.

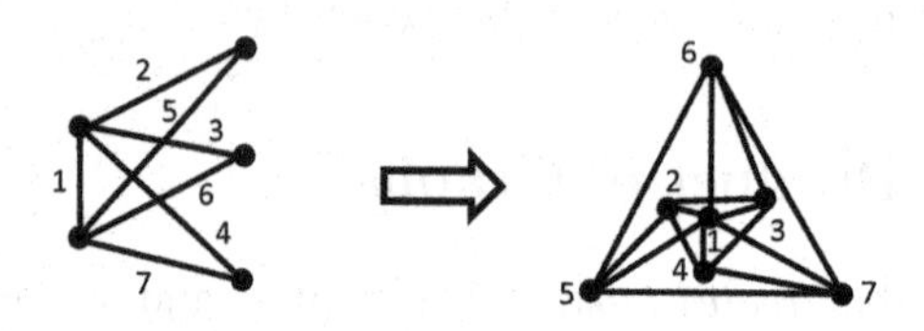

Figure 5.18: The line graph of a word-representable graph can be non-word-representable

Theorem 5.5.4 applied to even n, and Theorem 5.5.5 provide (infinitely many) other examples of word-representable graphs which map to non-word-representable graphs after taking the line graph operation. On the other hand, Theorem 5.5.4 applied to odd n, and Theorem 5.5.6 applied to graphs $L^4(G)$ (taking line graph operation applied four times) for any graph G different from a path, a cycle or the claw graph $K_{1,3}$, provide (infinitely many) examples of non-word-representable graphs which map to non-word-representable graphs after taking the line graph operation.

5.5 Line Graphs and Word-Representability

Definition 5.5.1. A *morphism* φ is a mapping $\Sigma^* \to \Sigma^*$ that satisfies the property $\varphi(uv) = \varphi(u)\varphi(v)$ for all words u and v. Clearly, the morphism is completely defined by its action on the letters of the alphabet. The *erasing* of a set $\Sigma\backslash S$ of elements is a morphism $\epsilon_S : \Sigma^* \to \Sigma^*$ such that $\epsilon_S(a) = a$ if $a \in S$ and $\epsilon_S(a) = \varepsilon$ otherwise, where ε is the empty word.

Example 5.5.2. If $S = \{2, 4\}$, $\Sigma = \{1, 2, 3, 4\}$, and $u = 32143231$ then $\epsilon_S(u) = 242$.

Let u be a word uniformly representing a graph $G = (V, E)$, and $S = S_1 \cup S_2 \cup \{a\} \subseteq V$ be such that $a \notin S_1 \cup S_2$ and $S_1 \cap S_2 = \emptyset$. By Remark 3.2.15, we can think of u as a cyclic word (that is, u being written on a circle), and we use the following notations:

- "$\forall(a\, S_1\, S_2\, a)$" for the statement "Between every two consecutive occurrences of a in u, each element of $S_1 \cup S_2$ occurs once and each element of S_1 occurs before any element of S_2", and
- "$\exists(a\, S_1\, S_2\, a)$" for the statement "There exist at least two consecutive occurrences of a in u, such that each element of $S_1 \cup S_2$ occurs between them and each element of S_1 occurs before any element of S_2".

In the case when S_1 or S_2 is a singleton, say $\{x\}$, we simply write x in the just introduced notation. Note that $\forall(a\, S_1\, S_2\, a)$ implies $\exists(a\, S_1\, S_2\, a)$ and is contrary to $\exists(a\, S_2\, S_1\, a)$. The quantifiers in these statements operate on pairs of consecutive occurrences of a in all cyclic shifts of a given word-representant. This notation may be generalized to an arbitrary number of sets S_i with the same interpretation. The following proposition illustrates the use of the quantifiers notation.

Proposition 5.5.3. *([94]) Let a word u represent a graph $G = (V, E)$, and $a, b, c \in V$ and $ab, ac \in E$. Then*

- *if $bc \notin E$ then both of the statements $\exists(a\, b\, c\, a)$, $\exists(a\, c\, b\, a)$ are true for u, while*
- *if $bc \in E$ then exactly one of the statements $\forall(a\, b\, c\, a)$, $\forall(a\, c\, b\, a)$ is true for u.*

Proof. Suppose $bc \notin E$. Since a, b and a, c alternate in u, at least one of $\exists(a\, b\, c\, a)$, $\exists(a\, c\, b\, a)$ is true. If only one of them is true for u, then b, c alternate in it, which is a contradiction with $bc \notin E$. If $bc \in E$ then the statement follows immediately from Proposition 3.2.14, since a, b, c form a clique of size 3. □

5.5.1 Line Graphs of Wheels

Recall Definition 3.5.1 for the notion of the wheel graph W_n and Definition 2.2.27 for the notion of a line graph.

Theorem 5.5.4. *([93, 94]) The line graph $L(W_n)$ is not word-representable for each $n \geq 4$.*

Proof. Let us describe $L(W_n)$ first. Denote the edges of the big (external) cycle of the wheel W_n by $e_1, \ldots, e_n$ in consecutive order and internal edges that connect the

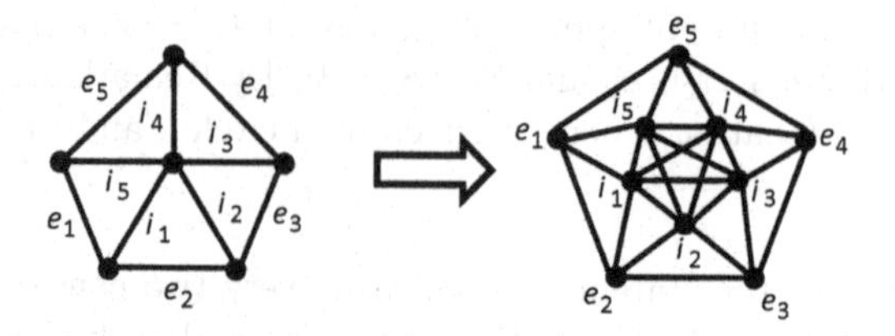

Figure 5.19: The wheel graph W_5 and its line graph $L(W_5)$

inside vertex to the big cycle by $i_1, \ldots, i_n$ so that an edge i_j is adjacent to e_j and e_{j+1} for $1 \leq j < n$, and i_n is adjacent to e_n and e_1. See the left graph in Figure 5.19 for the case $n = 5$.

In the line graph $L(W_n)$ the vertices $e_1, \ldots, e_n$ form a cycle where they occur consecutively, and the vertices $i_1, \ldots, i_n$ form a clique. In addition, vertices i_j are adjacent to e_j, e_{j+1} and i_n is adjacent to e_n, e_1. See the right graph in Figure 5.19 for the case $n = 5$.

Suppose that $L(W_n)$ can be represented by some word that we can assume to be uniform by Theorem 3.3.1. We will deduce a contradiction with Theorem 3.2.10.

Let $E = \{e_j : 1 \leqslant j \leqslant n\}$, $I = \{i_j : 1 \leqslant j \leqslant n\}$, and a uniform word $u = u_1u_2\cdots$ on the alphabet $E \cup I$ be a word-representant for $L(W_n)$. Due to Proposition 3.2.7, we can assume $u_1 = i_1$.

As we know from Proposition 3.2.14, the word $\epsilon_I(u)$ is of the form v^m, where $v = v_1 \cdots v_n$ is a permutation and $v_1 = i_1$. We will prove that v is either $i_1 \cdots i_n$ or $i_1 i_n i_{n-1} \cdots i_2$.

Suppose that there are some $\ell, k \in \{3, \ldots, n\}$ such that $\epsilon_{\{i_1,i_2,i_\ell,i_k\}}(v) = i_1 i_\ell i_2 i_k$. Note that $\ell \neq k$ due to v being a 1-uniform word. It is easy to see that the statement $\forall(i_1\, i_\ell\, i_2\, i_k\, i_1)$ is true for u. The vertex e_2 is adjacent neither to i_ℓ nor to i_k. By Proposition 5.5.3 this implies $\exists(i_1\, e_2\, i_\ell\, i_1)$ and $\exists(i_1\, i_k\, e_2\, i_1)$ are true for u. Taking into account the previous "for all" statement, we conclude that both $\exists(i_1\, e_2\, i_2\, i_1)$ and $\exists(i_1\, i_2\, e_2\, i_1)$ are true for u, which contradicts Proposition 5.5.3 applied to i_1, i_2, e_2. Thus, there are only two possible cases to consider, namely, $v = i_1 i_2 v_3 \cdots v_n$ and $v = i_1 v_2 \cdots v_{n-1} i_2$.

Using the same reasoning on a triple i_j, i_{j+1}, e_{j+1}, by induction on $j \geq 1$, we obtain $v = i_1 i_2 \cdots i_n$ for the first case and $v = i_1 i_n i_{n-1} \cdots i_2$ for the second one. It is sufficient to prove the theorem only for the first case, since reversing a uniform word and taking its cyclic shifts preserve the alternating properties.

By Proposition 5.5.3, exactly one of the statements $\forall(i_1\, e_1\, e_2\, i_1)$, $\forall(i_1\, e_2\, e_1\, i_1)$ is true for u. Let us prove that it is the statement $\forall(i_1\, e_1\, e_2\, i_1)$.

Applying Proposition 5.5.3 to the clique $\{i_1, i_2, e_2\}$ we have that exactly one of $\forall(i_1\, i_2\, e_2\, i_1)$, $\forall(i_1\, e_2\, i_2\, i_1)$ is true. Applying Proposition 5.5.3 to i_1, i_3, e_2 we have that both of $\exists(i_1\, e_2\, i_3\, i_1)$ and $\exists(i_1\, i_3\, e_2\, i_1)$ are true. The statement $\forall(i_1\, e_2\, i_2\, i_1)$ contradicts $\exists(i_1\, i_3\, e_2\, i_1)$ since we have $\forall(i_1\, i_2\, i_3\, i_1)$. Hence $\forall(i_1\, i_2\, e_2\, i_1)$ is true. Now applying Proposition 5.5.3 to i_1, e_1 and i_2 we have $\exists(i_1\, e_1\, i_2\, i_1)$. Taking into account $\forall(i_1\, i_2\, e_2\, i_1)$ and Proposition 5.5.3 applied to the clique $\{i_1, e_1, e_2\}$, we conclude that $\forall(i_1\, e_1\, e_2\, i_1)$ is true. In other words, between two consecutive i_1s in u there is an e_1 that occurs before an e_2.

Using the same reasoning, one can prove that the statement $\forall(i_n\, e_n\, e_1\, i_n)$ and the statements $\forall(i_j\, e_j\, e_{j+1}\, i_j)$ for each $j < n$ are true for u. We denote this set of statements by $(*)$.

The vertex e_1 is not adjacent to the vertex i_{n-1} but both of them are adjacent to i_1. Hence, by Proposition 5.5.3, somewhere in $\epsilon_{\{e_1, i_{n-1}, i_1\}}(u)$ the word $i_{n-1}e_1i_1$ occurs. Taking into account what we have already proved for v, this means that we have the structure $i_1 - i_2 - \cdots - i_{n-1} - e_1 - i_n - i_1 - i_2 - \cdots - e_n$ in u, where the elements of I do not occur in gaps denoted by "$-$".

Now inductively applying the statements $(*)$, we conclude that in u there is a structure $i_{n-1} - e_1 - e_2 - \cdots - e_n - i_{n-1}$ where no i_{n-1} occurs in the gaps. Suppose that e_1 occurs somewhere in the gaps between e_1 and e_n. Since e_1 and e_n are adjacent, that would mean that between two e_1s we have an occurrence of another e_n, which contradicts the fact that e_n and i_{n-1} are adjacent. One can prove that no element of E occurs in the gaps between e_1 and e_n in the structure we have found by using induction and arriving at a contradiction similar to the one above. In other words, $e_1e_2\cdots e_n$ occurs in the word $\epsilon_E(u)$ representing the cycle. This results in a contradiction with Theorem 3.2.10, which concludes the proof. □

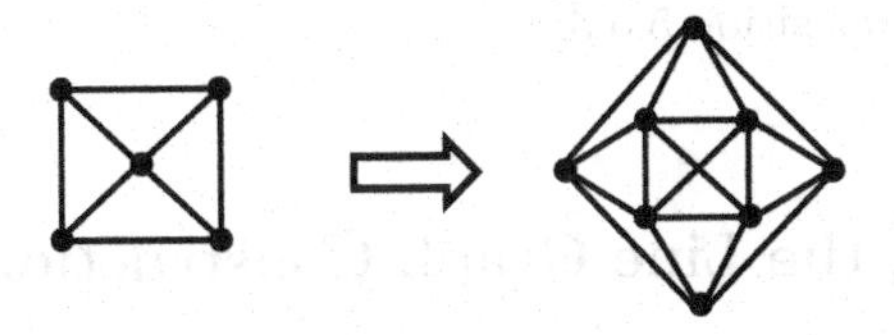

Figure 5.20: The wheel graph W_4 and its line graph $L(W_4)$

In Figure 5.20, we present the minimum graph, W_4, and its line graph to which Theorem 5.5.4 can be applied. Thus, the graph to the right, $L(W_4)$, in Figure 5.20 is not word-representable.

5.5.2 Line Graphs of Cliques

Theorem 5.5.5. *([93, 94]) The line graph $L(K_n)$ is not word-representable for each $n \geq 5$.*

Proof. It is not difficult to see that for any $n \geq 5$, $L(K_n)$ contains an induced subgraph $L(K_5)$. Thus, it is sufficient to prove the theorem for the case $n = 5$.

Let u be a word-representant for $L(K_5)$ with its vertices labelled as shown in Figure 5.21. Vertices $1, 2, a, b$ make a clique in $L(K_5)$. By applying Propositions 3.2.14 and 5.5.3 to this clique we see that exactly one of the following statements is true: $\forall(a\,\{1,2\}\,b\,a)$, $\forall(a\,b\,\{1,2\}\,a)$, $\forall(a\,x\,b\,\overline{x}\,a)$, where $x \in \{1,2\}$ and $\overline{x}$ is the negation of x, that is, $\overline{1} = 2$ and $\overline{2} = 1$. Because the reverse of u, $r(u)$, also represents $L(K_n)$ by Proposition 3.0.14, we only need to consider when either $\forall(a\,b\,\{1,2\}\,a)$ or $\forall(a\,x\,b\,\overline{x}\,a)$ is true.

(Case 1) Suppose that $\forall(a\,x\,b\,\overline{x}\,a)$ is true. The vertex 4 is adjacent to a, b, but not to 1, 2. Keeping in mind that a is also adjacent to 1 and 2, then applying Proposition 5.5.3 we have that both $\exists(a\,4\,\{1,2\}\,a)$ and $\exists(a\,\{1,2\}\,4\,a)$ are true. But between x, $\overline{x}$ there is b, so we have a contradiction $\exists(a\,4\,b\,a)$, $\exists(a\,b\,4\,a)$ with Proposition 5.5.3.

(Case 2a) Suppose that $\forall(a\,b\,1\,2\,a)$ is true. The vertex e is adjacent to a, 1, but not to b, 2. Applying Proposition 5.5.3 we have that both $\exists(a\,e\,b\,a)$ and $\exists(a\,2\,e\,a)$ are true. Taking into account the case condition, this implies that both $\exists(a\,e\,1\,a)$ and $\exists(a\,1\,e\,a)$ are true, which is a contradiction with Proposition 5.5.3.

(Case 2b) Suppose that $\forall(a\,b\,2\,1\,a)$ is true. The vertex 3 is adjacent to a, 2, but not to b, 1. Applying Proposition 5.5.3 we have $\exists(a\,3\,b\,a)$ and $\exists(a\,1\,3\,a)$. Again, taking into account the case condition this implies $\exists(a\,3\,2\,a)$ and $\exists(a\,2\,3\,a)$, which gives a contradiction with Proposition 5.5.3. □

5.5.3 Iterating the Line Graph Construction

It was shown by van Rooij and Wilf [133] that iterating the line graph operation on most graphs results in a sequence of graphs which grow (in the number of vertices) without bound. The exceptions are cycles, which stay as cycles of the same length, the claw graph $K_{1,3}$, which becomes a triangle after one iteration and then stays that way, and paths, which shrink to the empty graph (see Figure 5.22 for $K_{1,3}$ and examples of a cycle graph C_n and a path graph P_n). This unbounded growth results

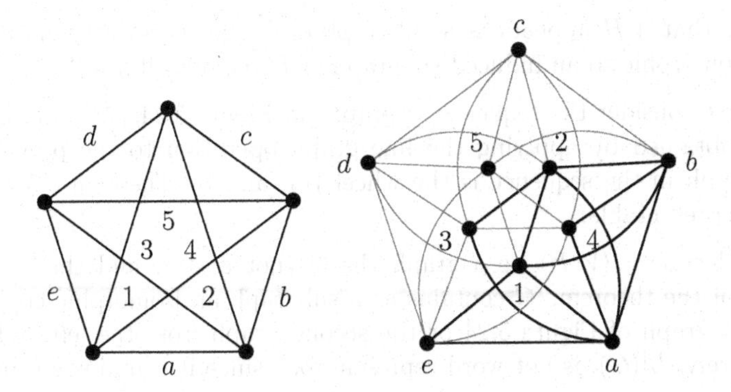

Figure 5.21: The complete graph K_5 and its line graph, where edges mentioned in the proof of Theorem 5.5.5 are drawn thicker

in graphs that are non-word-representable after a small number of iterations of the line graph operation since they contain the line graph of a large enough clique. A slight modification of this idea is used to prove the main result of [93, 94], which we record in Theorem 5.5.6 below. For the proof of this theorem, we need the notion of a *star graph* S_n, which is the complete bipartite graph $K_{1,n}$. Note that S_2 is the path graph P_3, and S_1 is the path graph P_2. Several star graphs are given in Figure 5.23.

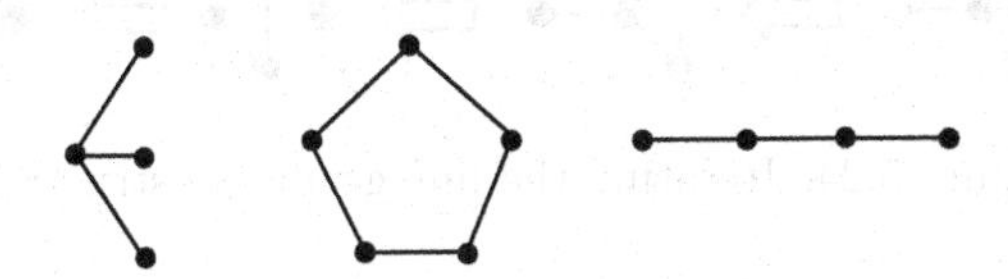

Figure 5.22: From left to right: graphs $K_{1,3}$, C_5 and P_4

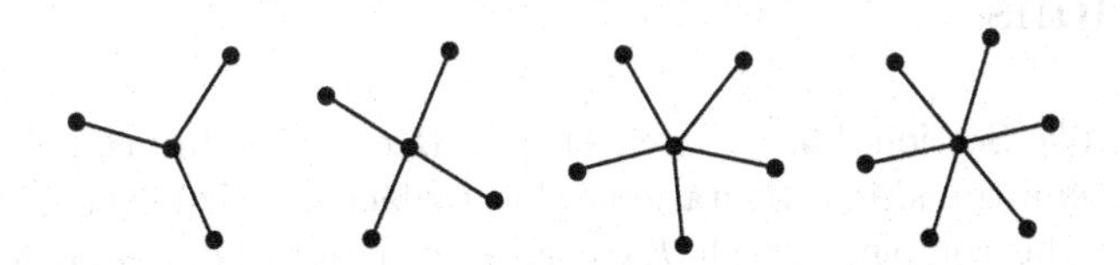

Figure 5.23: Star graphs S_3, S_4, S_5 and S_6

Theorem 5.5.6. *([93, 94]) If a connected graph G is not a path, a cycle or the claw graph $K_{1,3}$, then the line graph $L^n(G)$ is not word-representable for $n \geq 4$.*

Proof. Note that if H appears as a subgraph of G (not necessarily induced), then $L^n(H)$ is isomorphic to an induced subgraph of $L^n(G)$ for all $n \geqslant 1$.

We first consider the sequence of graphs in Figure 5.24. All but the leftmost graph are obtained by applying the line graph operation to the previous graph. The last graph in the sequence is the wheel W_4, and by Theorem 5.5.4, $L(W_4)$ is non-word-representable.

Now, let $G = (V, E)$ be a graph that is not a star, and that satisfies the conditions of the theorem. G contains as a subgraph an isomorphic copy of either the leftmost graph of Figure 5.24 or the second graph from the left. Thus $L^3(G)$ or, respectively, $L^4(G)$, is not word-representable, since it contains an induced line graph of W_4.

If G is a star S_n, where $n \geq 4$, then $L(G)$ is the complete graph K_n, and there is an isomorphic copy of the second from the left graph of Figure 5.24 in G, and $L^4(G)$ is not word-representable again. (Note that G is not S_1 or S_2 or S_3 because it is not a path or $K_{1,3}$.)

Finally, there is an isomorphic copy of the third graph of Figure 5.24 inside the fourth one. Therefore, the same reasoning can be used for $L^{4+k}(G)$ for each $k \geqslant 1$, which concludes the proof. □

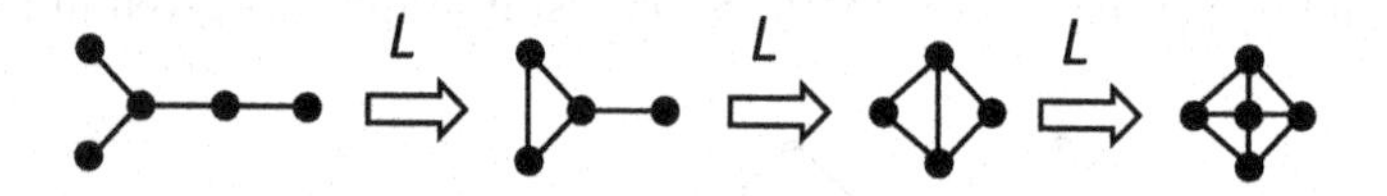

Figure 5.24: Iterating the line graph construction

5.6 Word-Representability of Polyomino Triangulations

This section, like Section 4.6, is based on [3]. Recall that in Section 4.6 we have shown, using semi-transitive orientations, that replacing each 4-cycle in an arbitrary polyomino by the complete graph K_4 results in a word-representable graph (see Theorem 4.6.5). Unlike that situation, triangulations of polyominoes are not always word-representable. However, there is an elegant result, Theorem 5.6.1 below, that shows that colourability determines word-representability in this case.

Recall the definition of a convex polyomino in Definition 4.6.3. Also, recall Theorem 4.3.3 saying that 3-colourable graphs are word-representable.

We will consider *triangulations* of a polyomino. Note that no triangulation is 2-colourable – at least three colours are needed to properly colour a triangulation, while four colours are always enough to colour any triangulation since we deal with planar graphs and it is well known that such graphs are 4-colourable. Not all triangulations of a polyomino are 3-colourable – for example, see Figure 5.25 for non-3-colourable triangulations (it is straightforward to check that they require four colours, and also that they are the only such triangulations, up to rotations, of a 3×3 grid graph). The main result in this section is the following theorem.

Theorem 5.6.1. *([3]) A triangulation of a convex polyomino is word-representable if and only if it is 3-colourable.*

In Subsection 5.6.1 we show that the two graphs in Figure 5.25 are non-word-representable. These graphs are to be used in the proof of Theorem 5.6.1 in Subsection 5.6.2. Also, in Subsection 5.6.2, we show that Theorem 5.6.1 is not true for an arbitrary polyomino.

5.6.1 Non-word-representability of Graphs in Figure 5.25

In the next theorem we show that the graphs T_1 and T_2 in Figure 5.25 are not word-representable.

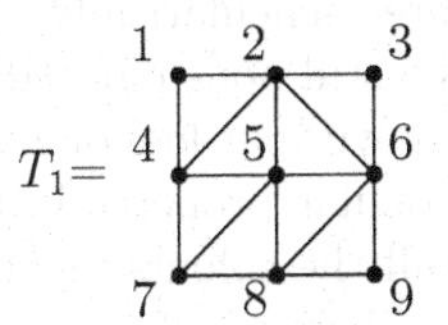

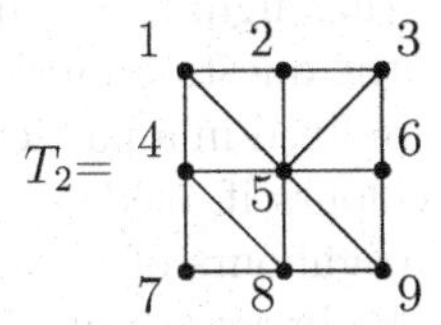

Figure 5.25: Non-3-colourable graphs T_1 and T_2

Theorem 5.6.2. *([3]) Graphs T_1 and T_2 in Figure 5.25 are not word-representable.*

Proof. Note that the graph T_1 contains as an induced subgraph the wheel W_5 formed by the vertices 2, 4, 5, 6, 7 and 8, and thus it is not word-representable by Proposition 3.5.3. On the other hand, the graph T_2 contains as an induced subgraph the wheel W_7 obtained by deleting the vertex 7, and thus it is not word-representable, again by Proposition 3.5.3. □

5.6.2 Triangulations of a Polyomino

In this subsection we first consider triangulations of a grid graph, and then generalize our arguments to triangulations of a convex polyomino by proving Theorem 5.6.1.

Then we show that Theorem 5.6.1 does not hold in the case of non-convex polyominoes.

Triangulations of a grid graph

Let $S = \{T_1, T_2\}$, where the graphs T_1 and T_2 are presented in Figure 5.25.

Lemma 5.6.3. *([3]) A triangulation T of a grid graph is 3-colourable if and only if it does not contain a graph from S as an induced subgraph.*

Proof. If T contains a graph from S as an induced subgraph, then it is obviously not 3-colourable.

For the opposite direction, suppose that T is not 3-colourable. We note that fixing colours of the leftmost top vertex in T and the vertex right below it uniquely determines the colours in the top two rows of T (a row is a horizontal path) if we use colours in $\{1, 2, 3\}$ and keep all other vertices of T uncoloured. We continue to colour all other vertices of T, row by row, from left to right using any of the available colours in $\{1, 2, 3\}$. At some point, colour 4 must be used (T is not 3-colourable). Let v be the first vertex coloured by 4. There are only three possible different situations when this can happen, which are presented in Figure 5.26 (numbers in this figure are colours). In that figure, the shaded area schematically indicates already coloured vertices of T, the question mark shows a still non-coloured vertex, and the colours adjacent to v are fixed in a particular way without loss of generality (we can rename already used colours if needed). A particular property in all cases is that among the colours of neighbours of v, we meet all the colours in $\{1, 2, 3\}$. Also, by our procedure, v must be in row i from the top, where $i \geq 3$.

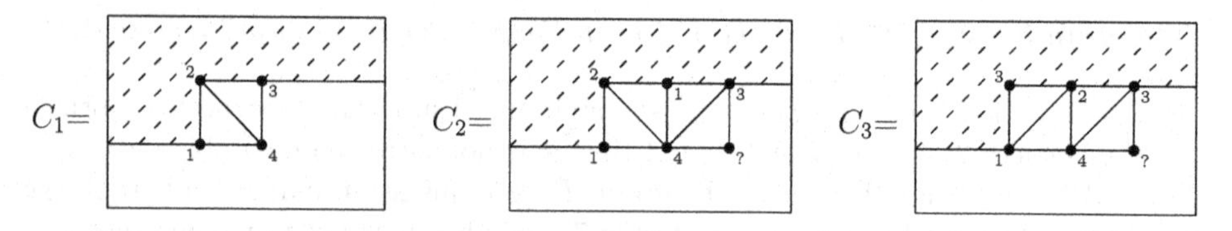

Figure 5.26: Three possible cases of appearance of colour 4 in colouring of T.

Case 1: Triangulation C_1 in Figure 5.26. Note that the vertex coloured by 1 must be in column i (from left to right), where $i \geq 2$; if the vertex were in column 1, there would be no need to colour it by 1 in our procedure — colour 3 could be used, contradicting the assumption (we would be in the conditions of Case 2 to be considered below). Thus, T must contain, as an induced subgraph, a 3×3 grid graph triangulation with the rightmost bottom vertex being v. We have four possible subcases depending on the colours of the vertices indicated by white circles in Figure 5.27, which allows us, in each case, to partially recover the triangulation of the 3×3 grid graph involved, as well as some of vertices' colours. However, in each of the cases, there is a unique way to complete the triangulation, namely, by joining the vertices coloured by 1 and 3. Indeed, if the vertex coloured by 2 is connected to a ? vertex, then the ? vertex must be coloured by 4 contradicting the fact that v was the first vertex coloured by 4 in our colouring procedure. We see that in each case, the triangulation belongs to S.

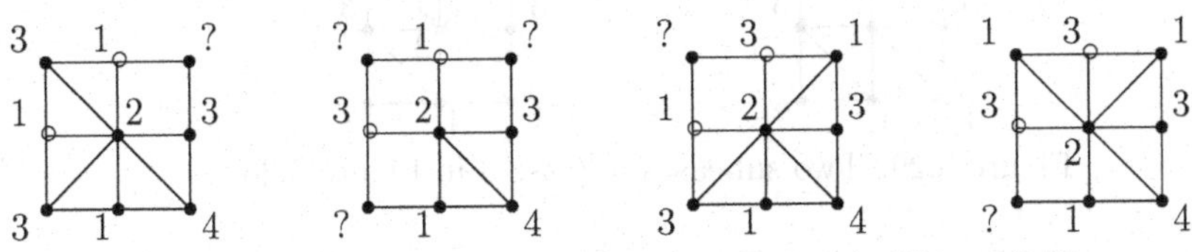

Figure 5.27: Four subcases in Case 1 in Figure 5.26

Case 2: Triangulation C_2 in Figure 5.26. In this case, T must contain, as an induced subgraph, a 3×3 grid graph triangulation with the bottom middle vertex being v. We have two possible subcases depending on the colour of the vertex indicated by a white circle in Figure 5.28, which allows us, in each case, to partially recover the triangulation of the 3×3 grid graph involved, as well as some of the vertices' colours. However, in each of the cases, there is a unique way to complete the triangulation, namely, by joining the vertices coloured by 2 and 3. Indeed, if the centre vertex coloured by 1 is connected to the ? vertex, then the ? vertex must be coloured by 4 contradicting the fact that v was the first vertex coloured by 4 in our colouring procedure. We see that in each case, the triangulation belongs to S.

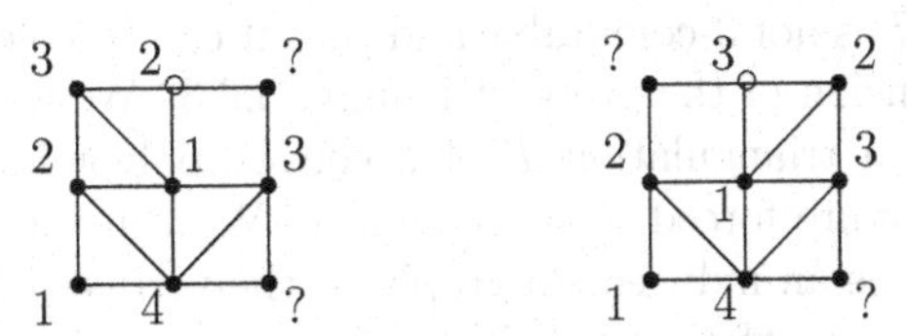

Figure 5.28: Two subcases in Case 2 in Figure 5.26

Case 3: Triangulation C_3 in Figure 5.26. In this case, similarly to Case 2, T must contain, as an induced subgraph, a 3×3 grid graph triangulation with the bottom middle vertex being v. We have two possible subcases depending on the colour of the vertex indicated by a white circle in Figure 5.29, which allows us in one case to partially recover the triangulation of the 3×3 grid graph involved, and in the other case to do so completely, obtaining as a result a triangulation in S. Also, we can recover some of the vertices' colours. We now see that in the partially recovered case, there is a unique way to complete the triangulation, namely, by joining the vertices coloured by 1 and 3. Indeed, if the centre vertex coloured by 2 is connected to one of the ? vertices, then the ? vertex must be coloured by 4, contradicting the fact that v was the first vertex coloured by 4 in our colouring procedure. We see that in any case, the triangulation belongs to S. □

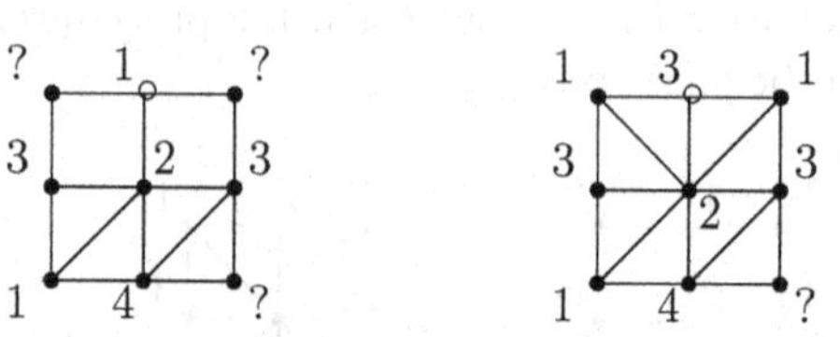

Figure 5.29: Two subcases in Case 3 in Figure 5.26

By Lemma 5.6.3 and Theorem 5.6.2 we have the truth of the following statement.

Theorem 5.6.4. *([3]) A triangulation of a grid graph is word-representable if and only if it is 3-colourable.*

Triangulations of a convex polyomino

Recall that $S = \{T_1, T_2\}$ defined in Figure 5.25.

Lemma 5.6.5. *([3]) A triangulation T of a convex polyomino is 3-colourable if and only if it does not contain a graph from S as an induced subgraph.*

Proof. Assume that T is not 3-colourable and thus it can be coloured in four colours. Our proof is an extension of the proof of Lemma 5.6.3. We use the same approach to colour vertices of a triangulation T of a convex polyomino as in the proof of Lemma 5.6.3 until we are forced to use colour 4. We will show that either T contains a graph from S as an induced subgraph, or the vertices coloured so far can be recoloured to avoid usage of colour 4; in the latter case our arguments can be repeated until eventually it will be shown that T contains a graph from S (otherwise a contradiction would be obtained with T being non-3-colourable). Once again, there are three possible situations, which are shown in Figure 5.30, where the areas of the convex polyomino labelled by A, B and C can possibly contain no other vertices than those shown in the figure coloured by 1, 2 and 3. We assume that vertices on the boundary of two areas belong to both areas; in particular, the vertex coloured by 2 in the leftmost picture in Figure 5.30 belongs to all three areas. Finally, we call a vertex in an area *internal* if it belongs only to a single area.

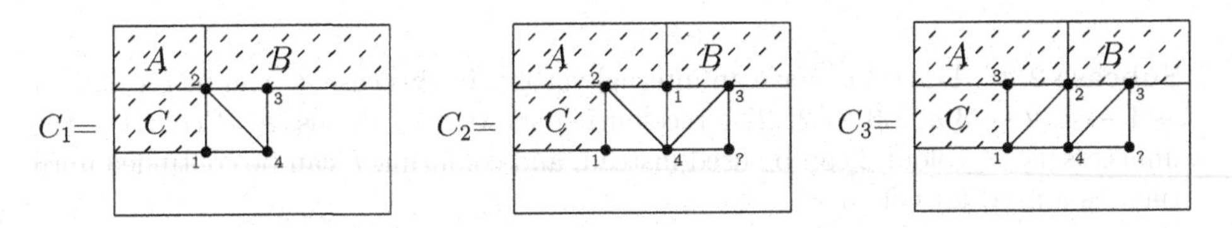

Figure 5.30: Three possible cases of appearance of colour 4 in colouring of T

Case 1: Triangulation C_1 in Figure 5.30. We consider two subcases here:

Subcase 1.1: A has at least one internal vertex. If B and C each have at least one internal vertex, then, taking into account that the polyomino is convex, T has a 3×3 grid graph with the bottom rightmost vertex coloured by 4 as an induced subgraph, and exactly the same arguments as in the proof of Case 1 in Lemma 5.6.3 can be applied to see that T contains a graph from S as an induced subgraph. On the other hand, if B (resp., C) does not have an internal vertex, then the vertex coloured by 3 (resp., 1) in the picture could be recoloured by 1 (resp., 3) so that there would be no need for colour 4, and we would continue colouring T until colour 4 needs to be used (then we would again find ourselves considering one of the three cases with more vertices already coloured).

Subcase 1.2: A does not have any internal vertices. We can recolour vertices of C as follows: $1 \to 3$, $2 \to 2$ and $3 \to 1$ (in particular, vertices coloured by 2 keep the same colour). Recolouring does not affect colouring in B, that is, we still have a proper colouring of a part of T. But then we see that usage of 4 is unnecessary: that vertex can be recoloured using colour 1, and we can continue colouring T until colour 4 needs to be used.

Case 2: Triangulation C_2 in Figure 5.30. Again, we consider two subcases here:

Subcase 2.1: A has at least one internal vertex. If B has an internal vertex then, taking into account convexity, we can use the argument in the proof of Case 2 in Lemma 5.6.3 applied to the 3×3 grid graph with the bottom rightmost vertex marked by ? to obtain the desired result. On the other hand, if B has no internal nodes, then the bottom border of B (containing at least two vertices coloured 1 and

3) can be recoloured as $1 \to 1$, $2 \to 3$ and $3 \to 2$ keeping the property of being a proper colouring. The recolouring shows that the usage of colour 4 was unnecessary — colour 3 can be used instead, and colouring T can be continued until there is a need for colour 4.

Subcase 2.2: A does not have an internal vertex. In this case, C can be recoloured as $1 \to 1$, $2 \to 3$ and $3 \to 2$. The recolouring shows that the usage of colour 4 was unnecessary — colour 2 can be used instead, and colouring T can be continued until there is a need for colour 4.

Case 3: Triangulation C_3 in Figure 5.30. Once again, we consider two subcases here:

Subcase 3.1: A has at least one internal vertex. If B has an internal vertex then, taking into account convexity, we can use the argument in the proof of Case 2 in Lemma 5.6.3 applied to the 3×3 grid graph with the bottom rightmost vertex marked by ? to obtain the desired result. On the other hand, if B has no internal nodes, then the bottom border of B (containing at least two vertices coloured 2 and 3) can be recoloured as $1 \to 3$, $2 \to 2$ and $3 \to 1$, keeping the property of being a proper colouring. The recolouring shows that the usage of colour 4 was unnecessary — colour 3 can be used instead, and colouring T can be continued until there is a need for colour 4.

Subcase 3.2: A does not have an internal vertex. In this final case, C can be recoloured as $1 \to 3$, $2 \to 2$ and $3 \to 1$. The recolouring shows that the usage of colour 4 was unnecessary — colour 1 can be used instead, and colouring T can be continued until there is a need for colour 4. □

The main result in this section, Theorem 5.6.1, now follows from Lemma 5.6.5 and Theorem 5.6.2.

Triangulations of an arbitrary polyomino

Theorem 5.6.1 is not true for triangulations of an arbitrary polyomino. Indeed, consider the triangulation T of the polyomino on seven squares in Figure 5.31 (where the centre square does not belong to the polyomino). It is easy to see that T is not 3-colourable, e.g. by letting without loss of generality the top leftmost vertex be coloured by 1 and the vertex horizontally next to it be coloured by 2, and to continuing using the least available colour for each vertex. On the other hand, T accepts the semi-transitive orientation shown to the right in Figure 5.31. This orientation is obtained by using the 4-colouring of T given in Figure 5.31 and orienting edges following the rules:

$$1 \to 2, 1 \to 3, 1 \to 4, 2 \to 3, 2 \to 4 \text{ and } 4 \to 3.$$

To see that the orientation is semi-transitive, one can observe that the only possible shortcuts must have a directed path $1 \to 2 \to 4 \to 3$ and the edge $1 \to 3$; however, there are only three directed paths $1 \to 2 \to 4 \to 3$ in the orientation, and in each case the beginning and end of such a path are not connected by an edge $1 \to 3$.

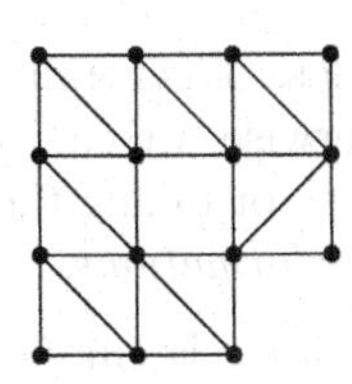

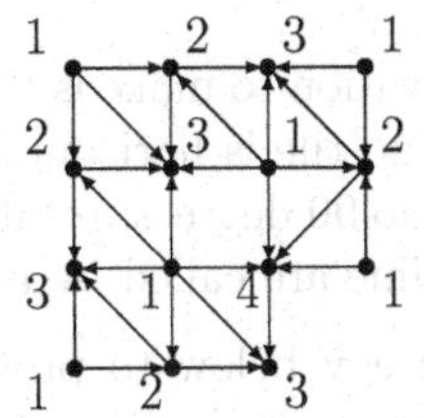

Figure 5.31: A non-3-colourable triangulation of a polyomino, and its semi-transitive orientation

5.7 More on Word-Representability of Polyomino Triangulations

Inspired by Theorem 5.6.1, Glen and Kitaev [68] considered the following variation of the polyomino triangulation problem discussed in Section 5.6. Polyominoes are objects formed by 1×1 tiles. A generalization of such graphs is to allow domino (1×2 or 2×1) tiles to be present in polyominoes. These graphs are called *polyominoes with domino tiles*. The problem then is to characterize those triangulations of such graphs that are word-representable. See Figure 5.32 for an example of a polyomino (of rectangular shape) with domino tiles (to the left) and one of its triangulations (to the right).

In [68], triangulations of rectangular polyominoes with a single domino tile were considered. The main result of that paper is the following generalization of Theorem 5.6.4.

Theorem 5.7.1. *([68]) A triangulation of a rectangular polyomino with a single domino tile is word-representable if and only if it is 3-colourable.*

The rest of the section is dedicated to sketching the proof of Theorem 5.7.1 presented in [68].

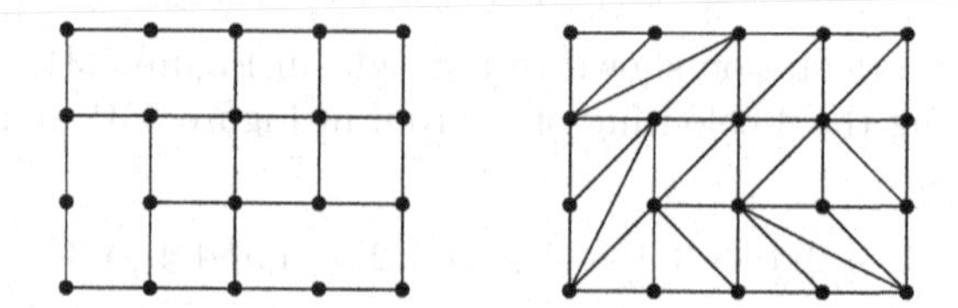

Figure 5.32: A polyomino with domino tiles (to the left) and one of its triangulations (to the right)

The first observation to make is that without loss of generality, we can assume that the single domino tile is horizontal, since otherwise, we can always rotate our rectangular polyomino 90 degrees; rotation of a shape or taking the mirror image of it with respect to a line are called by us *trivial transformations.*

While the strategy below to prove Theorem 5.7.1 is similar to the proof of Theorem 5.6.4, we have to deal with many more cases, resulting in 12 (non-equivalent up to trivial transformations) non-3-colourable and non-word-representable minimal graphs (which include the graphs T_1 and T_2 in Figure 5.25) instead of just two. All these graphs, except for T_1 and T_2, are listed in Figure 5.33. Non-representability of the graphs in Figure 5.33 follows from Proposition 3.5.3 by observing that A_1 contains W_9, while A_2, A_3, A_6 and A_7 contain W_7; also A_4, A_5, A_8, B_1 and B_2 contain W_5.

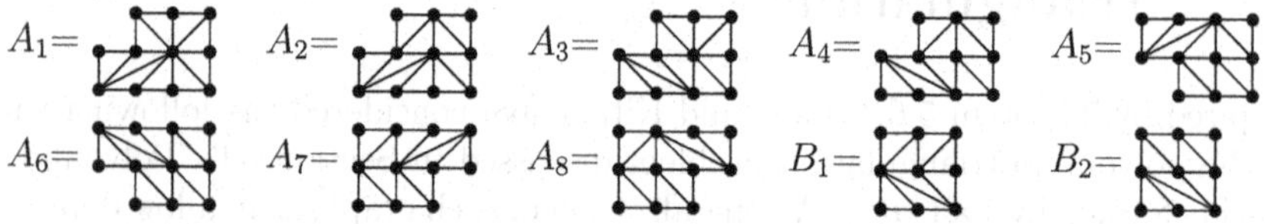

$A_1=$ $A_2=$ $A_3=$ $A_4=$ $A_5=$

$A_6=$ $A_7=$ $A_8=$ $B_1=$ $B_2=$

Figure 5.33: All minimal (non-equivalent up to trivial transformations) non-3-colourable and non-word-representable graphs (except for T_1 and T_2 in Figure 5.25) for triangulations of rectangular polyominoes with a single horizontal domino tile

Let M be the set of all graphs presented in Figures 5.25 and 5.33.

Lemma 5.7.2. *([68]) A triangulation T of a rectangular polyomino with a single domino tile is 3-colourable if and only if it does not contain a graph from M as an induced subgraph.*

Proof. If T contains a graph from M as an induced subgraph, then it is obviously not 3-colourable.

For the opposite direction, suppose that T is not 3-colourable. We note that fixing the colours of the leftmost top vertex in T and the vertex immediately below it uniquely determines the colours in the top two rows of T (a row is a horizontal path) if we are to use colours in $\{1,2,3\}$ and keep all other vertices of T uncoloured. We continue to colour all other vertices of T, row by row, from left to right using any of the available colours in $\{1,2,3\}$. At some point, colour 4 must be used (T is not 3-colourable) to colour, say, vertex v.

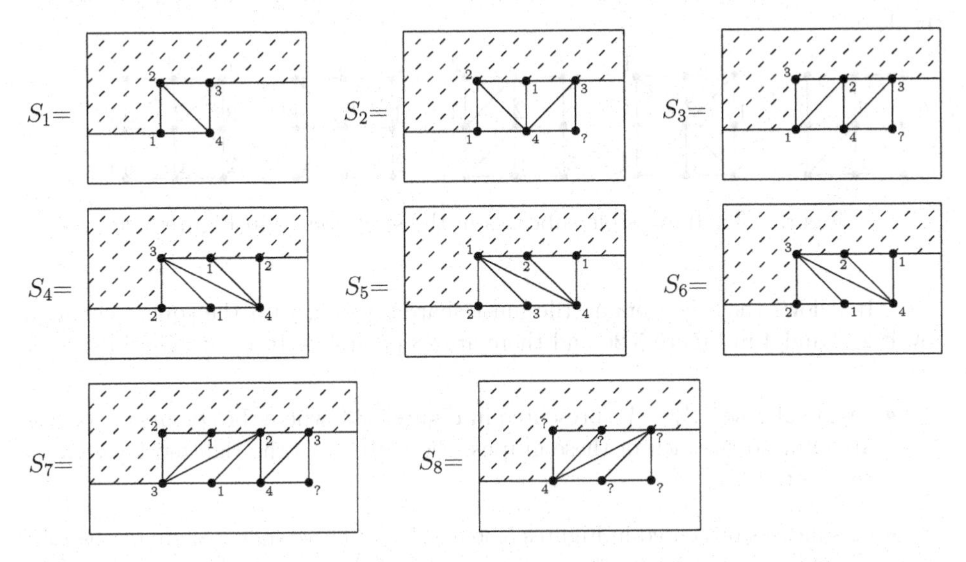

Figure 5.34: Eight possible cases of appearance of colour 4 in colouring of T

There are only eight possible different situations in which this can happen, which are presented in Figure 5.34 (numbers in this figure are colours). In that figure, the shaded area indicates schematically already coloured vertices of T, the question mark shows a still non-coloured vertex, and the colours adjacent to v are fixed in a particular way without loss of generality (we can rename already used colours if needed). A particular property in all cases is that among the colours of neighbours of v, we meet all the colours in $\{1,2,3\}$. Also, by our procedure, v must be in row i from the top, where $i \geq 3$, since the first two rows of any triangulation in question can be coloured in three colours.

Situation S_1. We can assume that the vertices coloured 1 and 2 are not in the leftmost coloumn, because otherwise instead of colour 1 we could use colour 3, and there would be no need to use colour 4 for colouring v. Further, note that in the case when the vertex v is involved in a subgraph presented schematically to the left in Figure 5.35 (the question marks there indicate that triangulations of respective squares are unknown to us), such a subgraph must be either T_1 or T_2. Indeed, otherwise, the subgraph must be one of the four graphs presented in Figure 5.35 to the right of the leftmost graph. However, in each of the four cases, we have a vertex labelled by * that would require colour 4, contradicting the fact that v is supposed to be the only vertex coloured by 4. This completes our considerations of eight of subcases in the situation S_1 out of 36. The remaining subcases are to be considered next.

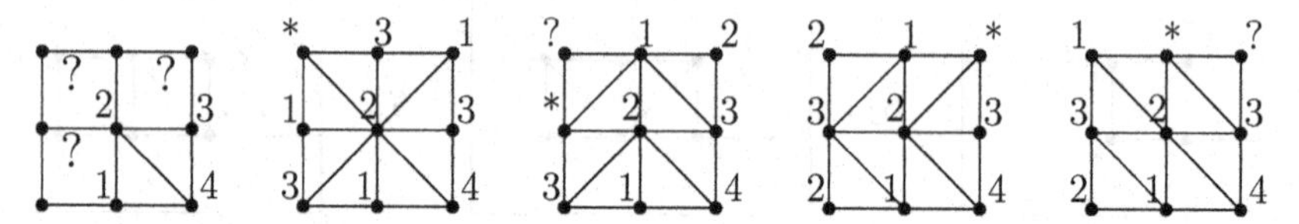

Figure 5.35: Impossible subcases in the situation S_1 in Figure 5.26

It follows that the domino tile must share a vertex with the square coloured by 1, 2, 3 and 4 in Figure 5.26 and there are 28 possible subcases to consider:

- Eight subcases, A_1—A_8, presented in Figure 5.33, where the colours of vertices are omitted (in each of these graphs, the vertex v is the rightmost vertex on the bottom row);
- 12 subcases presented in Figures 5.36 and 5.37. These subcases are impossible, because in each of them there exists a vertex, labeled by *, that requires colour 4;
- Four subcases presented in Figure 5.38, where colouring of vertices is shown; and, finally,
- Four subcases presented in Figure 5.39, where colouring of vertices of interest is shown.

However, the graphs in Figure 5.38, from left to right, are, respectively:

- A_3 flipped with respect to a horizontal line;
- A_1 flipped with respect to a horizontal line;
- A_1 rotated 180 degrees;

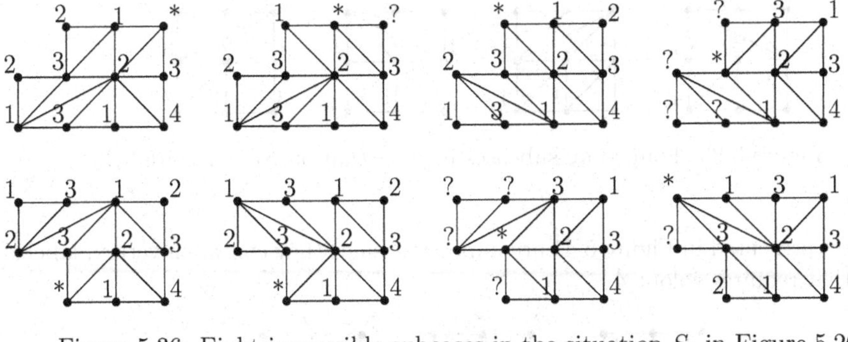

Figure 5.36: Eight impossible subcases in the situation S_1 in Figure 5.26

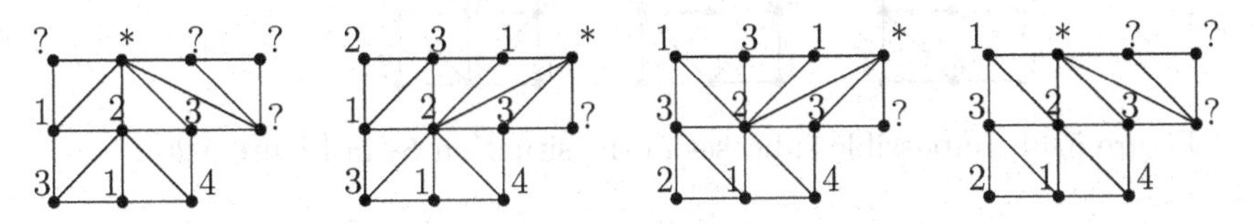

Figure 5.37: Four impossible subcases in the situation S_1 in Figure 5.26

- A_3 rotated 180 degrees.

Moreover, the leftmost two graphs in Figure 5.39 are, respectively, B_2 rotated 180 degrees and B_1 flipped with respect to a horizontal line. Finally, in each of the rightmost two graphs in Figure 5.39, we have a vertex labeled by * that would require colour 4 contradicting the fact that v is supposed to be the only vertex coloured by 4.

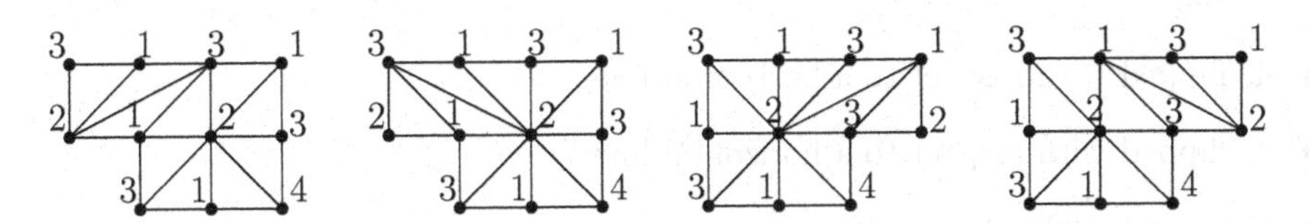

Figure 5.38: Four subcases in the situation S_1 in Figure 5.26

Situation S_2. Note that in the case when the vertex v is involved in a subgraph presented schematically to the left in Figure 5.40 (the question marks there indicate that triangulations of respective squares are unknown to us), such a subgraph must be either T_1 or T_2. Indeed, otherwise, the subgraph must be one of the two graphs presented in Figure 5.40 to the right of the leftmost graph. However, in each of the two cases, we have a vertex labeled by * that would require colour 4 contradicting the fact that v is supposed to be the only vertex coloured by 4. Similarly, the

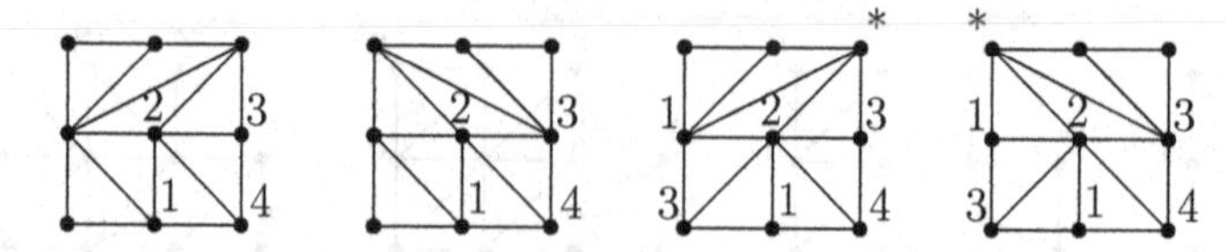

Figure 5.39: Four more subcases in the situation S_1 in Figure 5.26

subcases presented in Figure 5.41 are impossible since they contain a vertex, labeled by *, that requires colour 4.

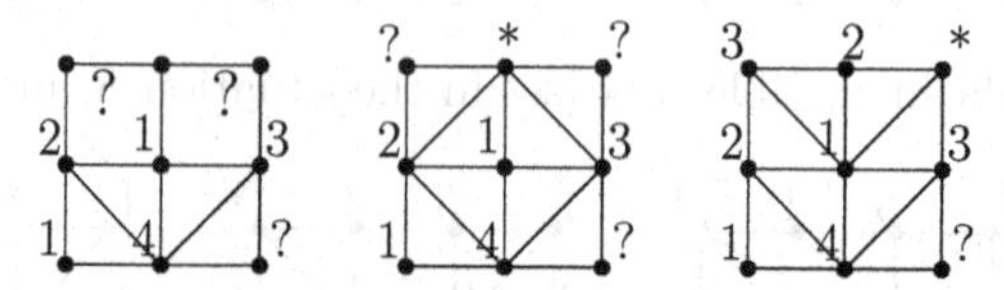

Figure 5.40: Impossible subcases in the situation S_2 in Figure 5.26

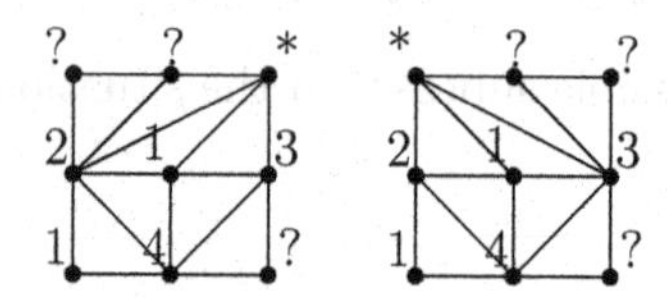

Figure 5.41: Two more impossible subcases in the situation S_2 in Figure 5.26

But then we have four possible subcases to be considered, which are presented in Figure 5.42, where colouring of vertices is shown. However, the graphs in Figure 5.42, from left to right, are, respectively:

- A_4 flipped with respect to a horizontal line;
- A_2 flipped with respect to a horizontal line;
- A_4 rotated 180 degrees;
- A_2 rotated 180 degrees.

Situation S_3. Note that in the case when the vertex v is involved in the subgraph presented schematically to the left in Figure 5.43 (the question marks there indicate that triangulations of the respective squares are unknown to us), such a subgraph must be either T_1 or T_2. Indeed, otherwise, the subgraph must be one of the two graphs presented in Figure 5.43 to the right of the leftmost graph. However, in each of the two cases, we have a vertex labelled by * that would require colour 4 contradicting the fact that v is supposed to be the only vertex coloured by 4.

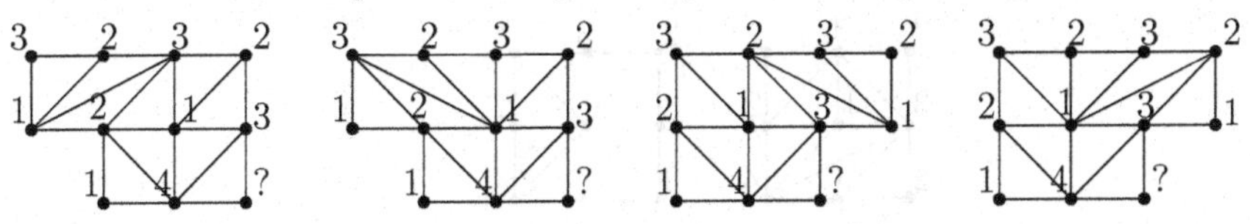

Figure 5.42: Four subcases in the situation S_2 in Figure 5.26

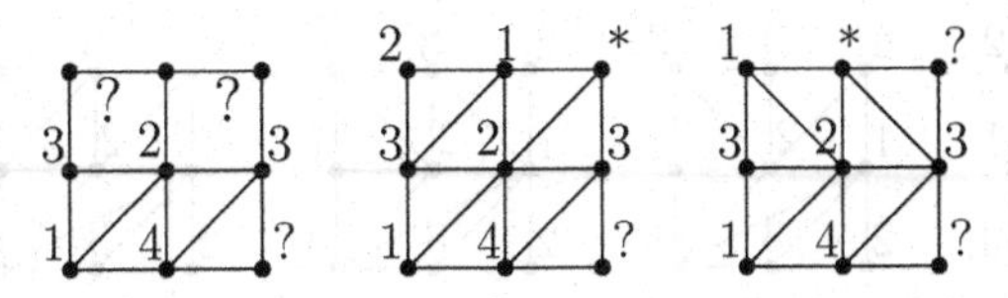

Figure 5.43: Impossible subcases in the situation S_3 in Figure 5.34

But then we have six possible subcases to consider: four subcases are presented in Figure 5.44, where colouring of vertices is shown, and two subcases correspond to B_1 rotated 180 degrees, and B_2 flipped with respect to a horizontal line; B_1 and B_2 are presented in Figure 5.33, where the colours of vertices are omitted (the vertex v corresponds to the middle vertex on the top row in each of the graphs). However, the graphs in Figure 5.44 are, respectively, A_7, A_8, A_6 and A_5 flipped with respect to a vertical line.

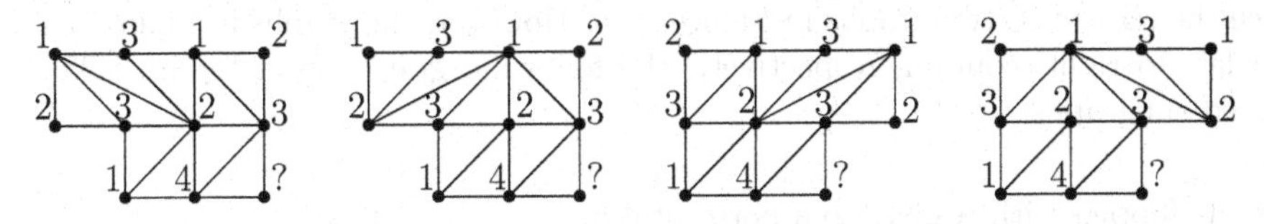

Figure 5.44: Four subcases in the situation S_3 in Figure 5.34

Situation S_4**.** In this case, we have only two subcases, namely B_1 and B_2 in Figure 5.33. Indeed, two other situations presented in Figure 5.45 are impossible (the vertices labelled by * there require colour 4, contradicting our choice of v).

Situation S_5**.** In this case, we have two possible and two impossible subcases, presented in Figure 5.46. The rightmost two graphs in that figure are impossible because the vertices labelled by * require colour 4. On the other hand, the leftmost two graphs are 3-colourable, which forces us to consider their extensions, namely larger subgraphs in the situation S_5.

Note that if there were no other vertices to the left of the leftmost two graphs in Figure 5.46, we could swap colours 2 and 3 in the bottom row to see that usage

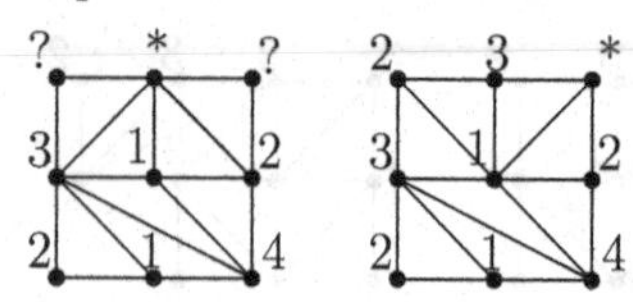

Figure 5.45: Impossible subcases in the situation S_4 in Figure 5.34

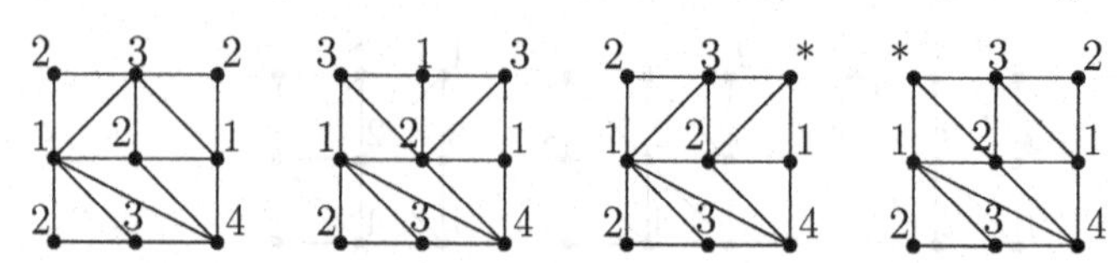

Figure 5.46: Possible and impossible subcases in the situation S_5 in Figure 5.34

of colour 4 for v is unnecessary. Thus, we can consider extensions of these graphs to the left. The leftmost graph in Figure 5.46 has two possible extensions, recorded as the two leftmost graphs in Figure 5.47 (colours are omitted in that figure), and two impossible extensions (because of the issue with using colour 4 more than once indicated by *) — see the leftmost two graphs in Figure 5.48. Finally, the next to leftmost graph in Figure 5.46 has two possible extensions, recorded as the two rightmost graphs in Figure 5.47 (colours are omitted in that figure), and two impossible extensions (because of the issue with using colour 4 more than once indicated by *) — see the rightmost two graphs in Figure 5.48. However, the graphs in Figure 5.47, from left to right, contain, respectively, the following graphs from Figure 5.33 as induced subgraphs:

- A_7 flipped with respect to a horizontal line;
- A_1 flipped with respect to a vertical line;
- A_2 flipped with respect to a vertical line;
- A_6 rotated 180 degrees.

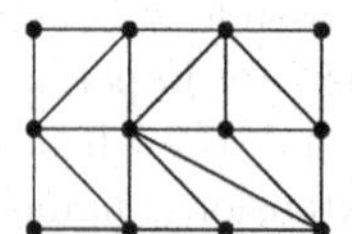
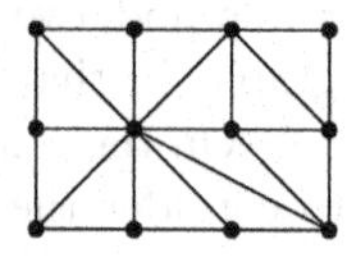
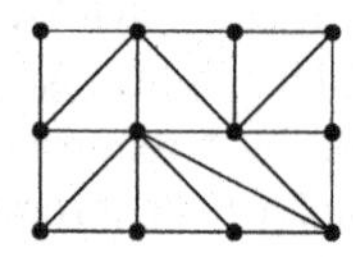
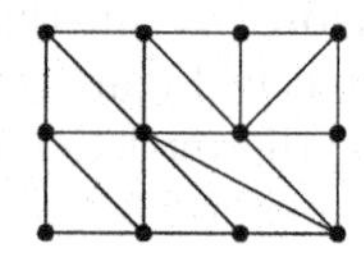

Figure 5.47: Possible extensions in the situation S_5 in Figure 5.34

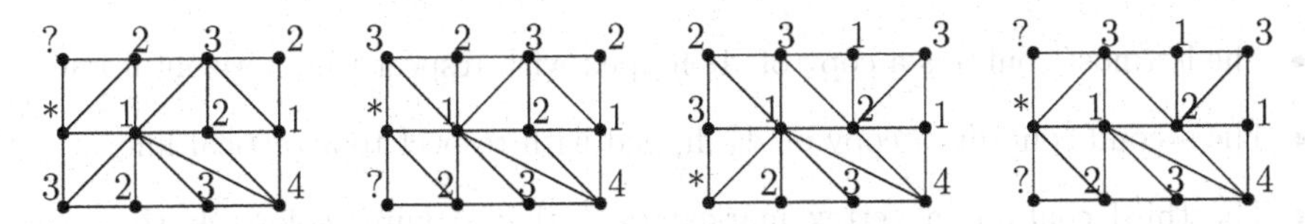

Figure 5.48: Impossible extensions in the situation S_5 in Figure 5.34.

Situation S_6. This situation is the same as situation S_4, since in both cases we have the same graph with three different colours in the top row.

Situation S_7. Note that the subcases in Figure 5.49 are not possible in this case, because the vertices labelled by * require usage of colour 4, contradicting the choice of the vertex v. Thus, in this situation, we only have two subcases, which are obtained from A_6 and A_7 in Figure 5.33, respectively, by flipping with respect to a horizontal line, and rotating 180 degrees.

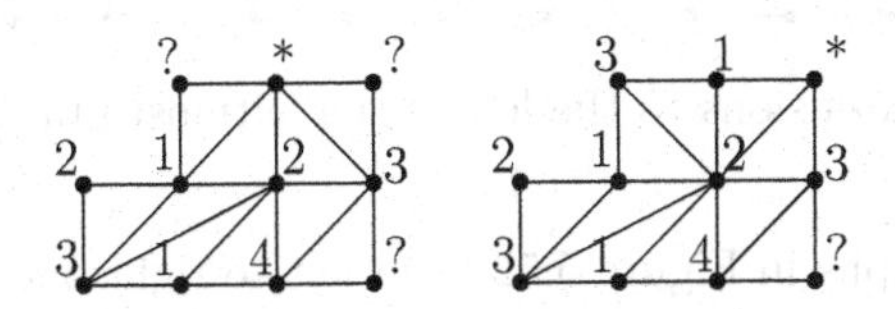

Figure 5.49: Impossible subcases in the situation S_7 in Figure 5.34

Situation S_8. We either have a copy of B_1 or B_2 flipped with respect to a vertical line, or we have one of the two subcases presented in Figure 5.50. Note that we can assume in Figure 5.50 that the vertices coloured by 1 (without loss of generality) are indeed of the same colour, since otherwise we would have a situation similar to that in Figure 5.45, which is impossible. However, this means that the vertex v coloured by 4 is not in the leftmost column, and we can consider eight subcases in which we extend extend the graphs in Figure 5.50 to the left: four extensions of the leftmost (resp., rightmost) graph are presented in Figure 5.51 (resp., Figure 5.52).

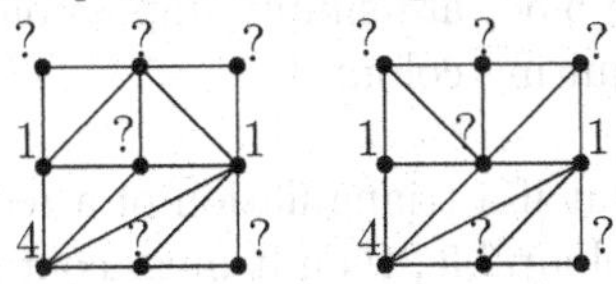

Figure 5.50: Subcases of interest in the situation S_8 in Figure 5.34

Consider the four graphs in Figure 5.51 from left to right one by one:

- The leftmost contains a copy of A_8 flipped with respect to a horizontal line.
- The second contains a copy of A_3 flipped with respect to a vertical line.
- The third contains a vertex marked by * that requires colour 4; thus this situation is impossible because of our choice of the vertex v.
- In the righthand graph, if the bottom leftmost vertex coloured by 2 were in the leftmost column, we could colour it by 1, and usage of colour 4 for colouring v would be unnecessary. Thus, the graph can be extended to the left, and out of four possible extensions, two contain (rotated) copies of T_1 or T_2, while the other two, presented in Figure 5.53, are simply impossible, because they contain a vertex, marked by *, requiring colour 4.

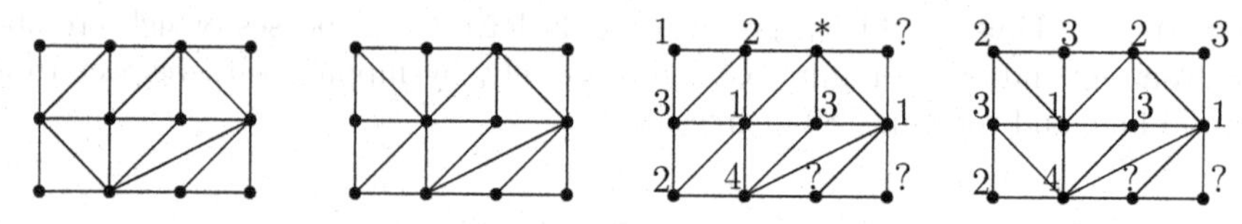

Figure 5.51: All possible extensions to the left of the leftmost graph in Figure 5.50

Consider the four graphs in Figure 5.52 from left to right one by one:

- The leftmost contains a copy of A_4 flipped with respect to a vertical line.
- The second contains a copy of A_5 rotated 180 degrees.
- The third contains a vertex marked by * that requires colour 4; thus this situation is impossible because of our choice of the vertex v.
- In the rightmost graph, if the bottom leftmost vertex coloured by 2 were in the leftmost column, we could colour it by 1, and usage of colour 4 for colouring v would be unnecessary. Thus, the graph can be extended to the left, and out of four possible extensions, two contain (rotated) copies of T_2, while the other two, presented in Figure 5.54, are simply impossible, because they contain a vertex, marked by *, requiring colour 4.

Thus, we have proved that if a triangulation of a rectangular polyomino with a single domino tile is not 3-colourable, then it must contain a graph from M as an induced subgraph. □

Theorem 5.7.1 now follows from Lemma 5.7.2 taking into account the fact that all graphs in M are non-word-representable.

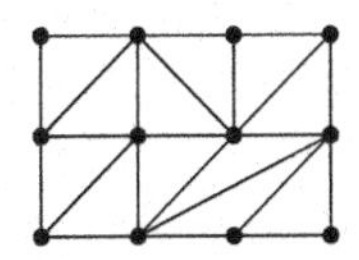

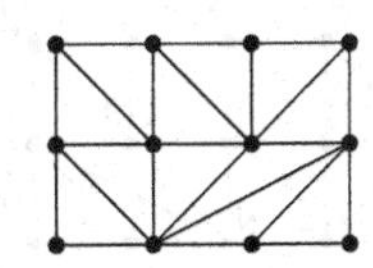

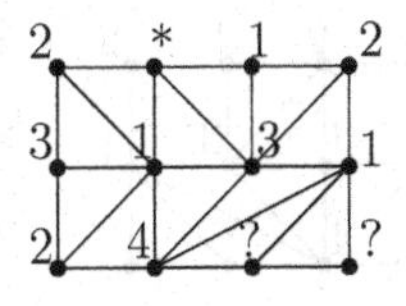

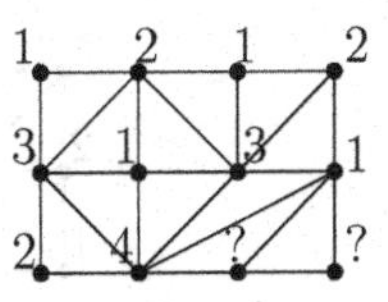

Figure 5.52: All possible extensions to the left of the rightmost graph in Figure 5.50

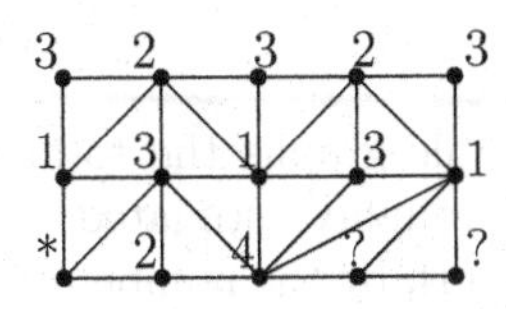

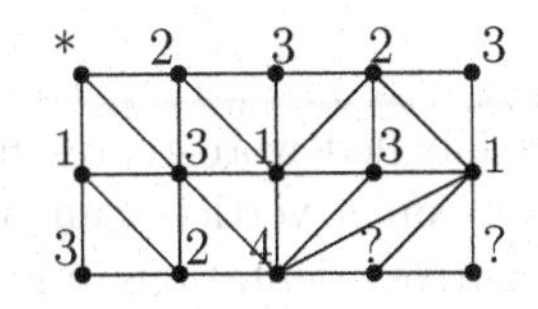

Figure 5.53: Possible T_1, T_2-avoiding extensions to the left of the rightmost graph in Figure 5.51

5.8 On the Number of Semi-transitive Orientations of Labelled Word-Representable Graphs

Clearly, any acyclic orientation of a complete graph K_n is semi-transitive, and thus there are $n!$ semi-transitive orientations of K_n in the case of labelled graphs. On the other hand, an edgeless graph on n vertices has exactly one semi-transitive orientation.

The goal of this section is to shed light on the total number of semi-transitive orientations of labelled word-representable graphs. To this end, we will employ a result on alternation word digraphs defined in Section 3.6 and the asymptotic enumeration of word-representable graphs discussed in Section 5.3.

It is known [125] that if c_n is the number of distinct alternation word digraphs for words over the alphabet $\{1, 2, \ldots, n\}$, one has $(\log_2 c_n) \in O(n^3)$.

As is mentioned in Section 3.6, it is straightforward from the definition of an alternation word digraph that taking its induced subgraph on the level of singletons (sets containing a single letter) and removing the orientation of all the edges, we will obtain a word-representable graph. Thus, the orientation on singletons we have just removed must be semi-transitive and c_n gives us an upper bound on the total number b_n of semi-transitive orientations of labelled word-representable graphs on n vertices (each alternation word digraph induces a semi-transitive orientation on the level of singletons). In particular, a rough upper bound for the average number of semi-transitive orientations of a labelled word-representable graph on n vertices is c_n/b_n, which is asymptotically $2^{O(n^3)}$.

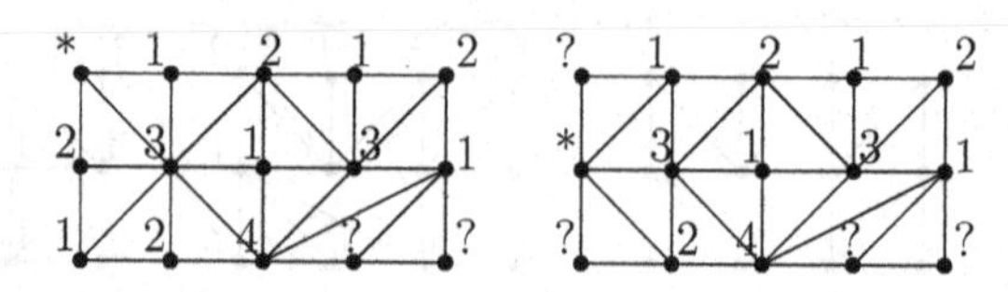

Figure 5.54: Possible T_2-avoiding extensions to the left of the rightmost graph in Figure 5.52

In the case of labelled word-representable graphs the total number of semi-transitive orientations on n vertices can be roughly estimated from below by the total number of transitive orientations, and thus, by the number of partially ordered sets on n elements, which is known (see [96]) to be $2^{\frac{n^2}{4}+\frac{3n}{2}+O(\log_2 n)}$. However, this number is less than b_n and therefore it is useless for estimating the average number in the labelled case. Thus, we only can give a trivial lower bound of 1 for the average number of semi-transitive orientations of a word-representable graph. In any case, we have the following theorem.

Theorem 5.8.1. *If b_n is the total number of semi-transitive orientations of labelled word-representable graphs on n vertices, we have*

$$2^{O(n^2)} \leq b_n \leq 2^{O(n^3)}.$$

5.9 Conclusion

In this chapter we discussed various results on word-representable graphs without a common thread. In particular, we know a characterization of word-representable graphs with representation number at most 2, while no such characterization is known for graphs with representation number 3 or higher. On the other hand, there are a number of nice results about graphs with representation number 3, probably the most notable one saying that for every, possibly non-word-representable, graph G there exists a 3-word-representable graph that contains G as a minor. We also know that the class of graphs with representation number 3 is not included in a class of c-colourable graphs for some constant c.

The chapter also offers a proof that asymptotically the number of word-representable graphs is $2^{\frac{n^2}{3}+o(n^2)}$. This is the only enumerative result on word-representable graphs known so far.

It turns out that some operations on word-representable graphs are safe, in the sense that applying such operations to word-representable graphs produces word-representable graphs, while some other operations are not safe. Safe operations include gluing two graphs at a vertex, connecting two graphs by an edge, replacing a vertex with a module, Cartesian product, and rooted product. Unsafe operations include taking graph's complement, edge contraction, gluing two graphs at an edge and taking the line graph.

Speaking of line graphs, it is remarkable that if a connected graph G is not a path, a cycle or the claw graph $K_{1,3}$, then iterating the taking the line graph operation starting from G four or more times will necessarily result in a non-word-representable graph.

Some results give an elegant characterization of word-representability. An example of such a result is the following theorem: a triangulation of a convex polyomino is word-representable if and only if it is 3-colourable.

5.10 Exercises

1. Label the graph in Figure 5.55. Show that this graph is a circle graph by representing its vertices by inscribed chords with the right intersection properties. Based on this, find a 2-word-representation of the graph.

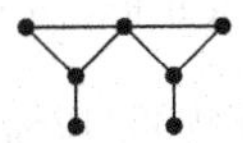

Figure 5.55: The graph for Exercise 1

2. Recall a 3-word-representation of the Petersen graph in Figure 5.3. Then use the proof of Theorem 5.2.2 to construct a 3-word-representation of the graph obtained from the Petersen graph by adding to it the path $1, a, b, y, 3$ of length 4.

3. Using Theorem 5.2.8, for each non-word-representable graph G in Figure 3.6, provide an example of a word-representable graph H that contains G as a minor.

4. Extend Table 5.1 by providing a 3-representation of the graphs L_5, L_6, Pr_5 and Pr_6.

5. Use Table 5.1 and Theorem 5.4.7 to 3-represent the graph to the right in Figure 5.14.

6. Use Table 5.1 and the construction in the proof of Theorem 5.4.1 to 3-represent the graph obtained from the two leftmost prisms in Figure 5.6 by connecting their vertices 1 by an edge. Note that one needs to rename the vertices of one of the prisms.

7. Use Table 5.1 and the construction in the proof of Theorem 5.4.2 to 3-represent the graph obtained from the two leftmost prisms in Figure 5.6 by gluing them at the vertices 1. Note that one needs to rename the vertices of one of the prisms.

8. Show that the line graph $L(\mathrm{Pr}_3)$ of the prism graph Pr_3 is word-representable.

9. Suppose that G is the leftmost graph in Figure 3.1. Show that $L^3(G)$ is the graph to the right in Figure 5.20, and then conclude by Theorem 5.5.4 that $L^3(G)$ is not word-representable.

5.11 Solutions to Selected Exercises

2 Consider the first representation of the Petersen graph given in the proof of Theorem 5.2.1 and turn it into a representation of the graph G obtained from the Petersen graph by attaching a leaf y to the vertex 3 as follows:

$$1y3y87296(10)749354128y3(10)685(10)194562.$$

We will now use the proof of Theorem 5.2.2 to attach the path $1, a, b, y$ of length 3 to G. Note that the roles of x, u, v, y in that proof are played respectively by $1, a, b, y$, and we find ourselves in case 1 considered in the proof (described by the possibility $x^1y^1y^2x^2y^3x^3$). Thus, we need to substitute y^2 by $b^1a^1y^2b^2$ and 1^3 by $a^21^3b^3s^3$ to obtain the desired representation:

$$1y3bayb87296(10)749354128y3(10)685(10)a1ba94562.$$

5 According to Table 5.1, the prism Pr_3 to the left in Figure 5.14 can be represented by $313'231'12'23'1'2'33'11'22'$. But then, using the substitution in the proof of Theorem 5.4.7 , we see that the word $3abc3'231'abc2'23'1'2'33'abc1'22'$ 3-represents the graph in question.

Chapter 6

Representing Graphs via Pattern-Avoiding Words

Chapters 3–5 study the notion of word-representable graphs. This chapter, based on [84, 90], provides a far-reaching generalization of the notion of a word-representable graph, namely that of a *u-representable graph*. Section 7.7 contains other ways, including another generalization, to define the notion of word-representability.

A key observation made by Jeffrey Remmel in 2014 is that alternation of letters x and y in a word w (see Definition 3.0.3) is equivalent to *matching avoidance* of the *pattern* 11 in the subsequence formed by all occurrences of x and y in w. That is, in this subsequence xs and ys never stand next to each other. Thus, word-representable graphs are *11-representable graphs* in the terminology of this chapter. But what happens if we preplace the pattern 11 by the pattern 111? Or by 1111? Or, if the letters in w are ordered, by 12? Or by 121? Or by any other pattern over $\{1, 2\}$? We refer to Section 6.1 for precise definitions.

Taking into account how involved the theory of word-representable graphs is, it comes as an unexpected phenomenon that *any* (simple) graph is easily u-representable for any u of length at least 3 (see Section 6.2). In either case, the main focus of this chapter is on presenting results on 12-representable graphs. In particular, in Section 6.5, we will show that the class of these graphs is properly included in the class of comparability graphs, and it properly includes the classes of *co-interval graphs* and *permutation graphs* as shown in Figure 6.1. It turns out that the notion of 12-representable graphs is a natural generalization of the notion of permutation graphs.

Figure 6.1 also gives examples of graphs inside/outside the involved graph classes. For instance, even cycles of length at least 6, being comparability graphs, are not 12-representable (see Theorem 6.3.14 below); also, odd wheels on six or more vertices are not 11-representable by Proposition 3.5.3.

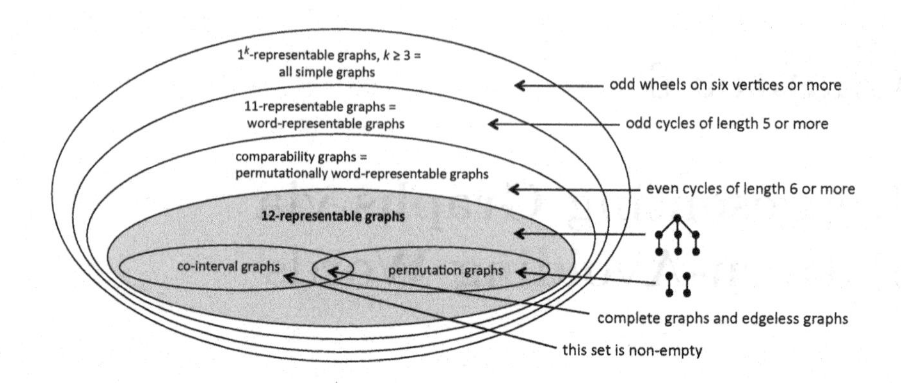

Figure 6.1: The place of 12-representable graphs in a hierarchy of graph classes

A fundamental difference between word-representable (i.e. 11-representable) graphs and 12-representable graphs (where 12 can be replaced by any pattern involving both 1s and 2s) is as follows. In the former case, unless dealing with enumeration, it is not so important whether we deal with labelled or unlabelled graphs: isomorphic graphs are either word-representable or not. On the other hand, in the latter case, by labelling the same graph differently we may witness both 12-representable and non-12-representable copies of the same graph. Thus, in our definition, an unlabelled graph is 12-representable if it admits a labelling giving a 12-representable labelled graph.

The chapter is organized as follows. In Section 6.1, we give all necessary definitions and some straightforward facts. In Section 6.2 we will show that any graph is u-representable assuming that u is of length at least 3. Some basic properties of 12-representable graphs are established in Section 6.3. In Section 6.4, we provide a characterization of 12-representable trees. Even though all trees are easily 11-representable (i.e. word-representable) using two copies of each letter (see Section 3.1), not all of them are 12-representable. It turns out that a tree is 12-representable if and only if it is a *double caterpillar* defined in Section 6.4. In Section 6.5, we compare the class of 12-representable graphs to other graph classes, thus explaining Figure 6.1. Finally, in Section 6.6, we discuss 12-representability of induced subgraphs of a grid graph.

6.1 From Letter Alternations to Pattern Avoidance

Let $\mathbb{P} = \{1, 2, \ldots\}$ denote the set of positive integers and $\mathbb{P}^*$ the set of all words over $\mathbb{P}$. Recall from Definition 3.0.1 that for a word $w = w_1 \cdots w_n$ in $\mathbb{P}^*$, $A(w)$ denotes the set of letters occurring in w.

Definition 6.1.1. If $B \subseteq A(w)$, then we let w_B be the word that results from w by removing all the letters in $A(w) \backslash B$.

Example 6.1.2. If $w = 4513113458$, then $w_{\{1,3,5\}} = 5131135$.

Definition 6.1.3. Given a word $u \in \mathbb{P}^*$, we let $\text{red}(u)$ be the word that is obtained from u by replacing each occurrence of the ith smallest letter that occurs in u by i.

Example 6.1.4. If $u = 347439$, then $\text{red}(u) = 123214$.

Definition 6.1.5. Given a word $u = u_1 \cdots u_j \in \mathbb{P}^*$ such that $\text{red}(u) = u$, and a word $w = w_1 \cdots w_n \in \mathbb{P}^*$, we say that w has a *u-match starting at position i* if $\text{red}(w_i w_{i+1} \cdots w_{i+j-1}) = u$.

Example 6.1.6. The word 23144522146 has 12-matches starting at positions 1, 3, 5, 9 and 10.

The following definition gives an analogue of the function red on graphs.

Definition 6.1.7. If $G = (V, E)$ is a labelled graph such that $V \subset \mathbb{P}$ and $|V| = n$, then the *reduction* of G, denoted $\text{red}(G)$, is the graph $G' = (\{1, \ldots, n\}, E')$ that results by replacing the ith smallest label of V by i.

Example 6.1.8. See Figure 6.2 for an example of graph reduction.

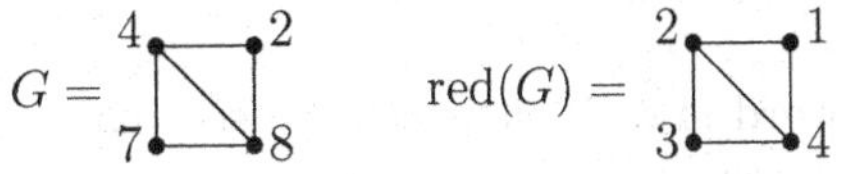

Figure 6.2: Reduction of a graph

The key to a generalization of the notion of a word-representable graph is to reframe this notion in the language of patterns in words. Note that x and y alternate in a word w if and only if $w_{\{x,y\}}$ has no 11-match. Thus, a graph $G = (V, E)$ is word-representable if and only if there is a word $w \in \mathbb{P}^*$ such that $A(w) = V$ and for all $x, y \in V$, xy is not an edge in E if and only if $w_{\{x,y\}}$ has a 11-match. This leads us to the following definition.

Definition 6.1.9. Let $u = u_1 \cdots u_j$ be a word in $\{1,2\}^*$ such that $\text{red}(u) = u$. Then we say that a labelled graph $G = (V,E)$, where $V \subset \mathbb{P}$, is *u-representable* if there is a word $w \in \mathbb{P}^*$ such that $A(w) = V$ and for all $x, y \in V$, $xy \notin E$ if and only if $w_{\{x,y\}}$ has a u-match. An unlabelled graph H is u-representable if it admits a labelling resulting in a u-representable labelled graph H'. We say that H' *realizes u-representability* of H, and we assume that labels in H' take all values in $\{1,\ldots,n\}$ for some n.

Thus, by Definition 6.1.9, G is word-representable if and only if G is 11-representable. Note that by replacing "word-representable graphs" by "u-representable graphs" in Proposition 3.0.8, we obtain a truth statement establishing the hereditary nature of u-representable graphs.

Definition 6.1.10. Suppose that $w = w_1 w_2 \cdots w_j \in \{1,\ldots,n\}^*$ and $A(w) = \{1,\ldots,n\}$. The *complement* of w is the word

$$c(w) = (n+1-w_1)(n+1-w_2)\cdots(n+1-w_j).$$

Example 6.1.11. For the word $w = 3552314$, the complement $c(w) = 3114352$.

There are some natural symmetries among u-representable graphs. That is, suppose that $u = u_1 \cdots u_j \in \{1,\ldots,n\}^*$ and $A(u) = \{1,\ldots,n\}$. This ensures that $\text{red}(u) = u$. Recall that the reverse of u is the word $r(u) = u_j u_{j-1} \cdots u_1$. Then for any word $w \in \mathbb{P}^*$, it is easy to see that w has a u-match if and only if $r(w)$ has a $r(u)$-match. We have the following statement generalizing Proposition 3.0.14.

Proposition 6.1.12. *([84]) Let $G = (V,E)$ be a graph and $u \in \mathbb{P}^*$ such that $\text{red}(u) = u$. Then G is u-representable if and only if G is $r(u)$-representable.*

Proof. It is easy to see that if w witnesses that G is u-representable then $r(w)$ witnesses that G is $r(u)$-representable. □

Similarly, for any word $w \in \mathbb{P}^*$, it is easy to see that w has a u-match if and only if $c(w)$ has a $c(u)$-match.

Definition 6.1.13. Given a graph $G = (\{1,\ldots,n\},E)$, we let the *supplement* of G, $c(G)$ be defined by $c(G) = (V, c(E))$, where for all $x, y \in V$, $xy \in E$ if and only if $(n+1-x)(n+1-y) \in c(E)$.

One can think of the supplement of the graph $G = (V,E)$ as replacing each label x on a vertex of G by the label $n+1-x$. For example, Figure 6.3 depicts a labelled graph on $\{1,2,3,4\}$ and its supplement. Then we have the following statement.

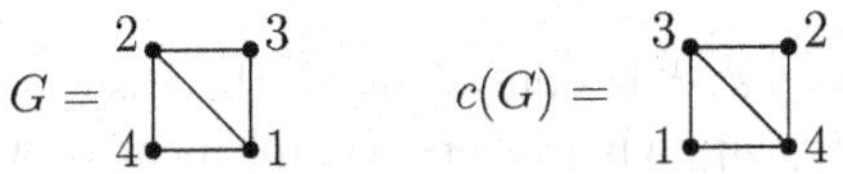

Figure 6.3: The supplement of a graph $G = (\{1, \ldots, n\}, E)$

Proposition 6.1.14. *([84]) Let $G = (V, E)$ be a graph, and u be a word in $\{1, \ldots, n\}^*$ such that $A(u) = \{1, \ldots, n\}$. Then G is u-representable if and only if $c(G)$ is $c(u)$-representable.*

Proof. It is easy to see that if w witnesses that G is u-representable, then $c(w)$ witnesses that $c(G)$ is $c(u)$-representable. □

As a direct corollary to Proposition 6.1.14 we have the following statement.

Corollary 6.1.15. *([84]) Let u be a word in $\{1, \ldots, n\}^*$ such that $A(u) = \{1, \ldots, n\}$. If a labelled graph $G' = (V', E')$ realizes u-representability of a graph G, and a vertex $v' \in V' = \{1, 2, \ldots, n\}$ has label 1 (resp., n), then there exists a labelled graph $G'' = (V', E'')$ realizing $c(u)$-representability of G where the vertex v' has label n (resp., 1).*

6.2 u-representable graphs for u of length 3 or more

The theory of word-representable graphs is rather involved, and thus this section material is rather surprising, since we will see that *any* graph can be u-represented assuming that u is of length 3 or more.

In what follows, for a letter x, x^k denotes k concatenated copies of x.

By Proposition 6.1.14, while studying u-representation of graphs, we can assume that $u = 1u'$ for some $u' \in \{1, 2\}^*$. Thus we only need to consider the following four disjoint cases for $u = 1u_2u_3 \cdots u_k$, where $k \geq 3$.

- $u = 1^k$, where $k \geq 3$ (Theorem 6.2.2),
- $u = 1^a2^b1u_{a+b+2}u_{a+b+3} \cdots u_k$, where $a, b \geq 1$ (Theorem 6.2.3),
- $u = 1^{k-1}2$, where $k \geq 3$ (Theorem 6.2.4),
- $u = 1^a2^b$, where $a, b \geq 2$ and $a + b = k$ (Theorem 6.2.5).

Note that the case $u = 12^{k-1}$ is equivalent to the case $u = 1^{k-1}2$ by Propositions 6.1.12 and 6.1.14 (by applying reverse complement to words u-representing a graph), and thus it is omitted. In each of the four cases above, we will show that any graph can be u-represented.

Lemma 6.2.1. *The complete graph K_n on a vertex set $\{1, \ldots, n\}$ is u-representable for every u of length at least* 3.

Proof. K_n can be represented by any permutation of $\{1, \ldots, n\}$, e.g. by $12 \cdots n$. □

Theorem 6.2.2. *([84]) Any graph $G = (V, E)$ is 1^k-representable for any $k \geq 3$.*

Proof. Fix any $k \geq 3$. Suppose that $V = \{1, \ldots, n\}$. If $G = K_n$ then by Lemma 6.2.1 G is 1^k-representable.

We proceed by induction on the number of edges in a graph with the base case being the complete graph. Our goal is to show that if G is 1^k-representable, then the graph G' obtained from G by removing any edge ij is also 1^k-representable.

Suppose that w 1^k-represents G. Recall the notion $p(w)$ of the initial permutation of w in Definition 3.2.11. That is, $p(w)$ is obtained from w by removing all but the leftmost occurrence of each letter. Let π be any permutation of $V \backslash \{i, j\}$. Then we claim that the word

$$w' = i^{k-1}\pi i p(w) w$$

1^k-represents G'. Indeed, the vertices i and j are not connected any more because $w'_{\{i,j\}}$ contains i^k. Also, no new edge can be created because of the presence of w as a subword. Thus, we only need to show that each edge is, $s \neq j$, represented by w is still represented by w', and each edge jm, $m \neq i$, represented by w is still represented by w'.

Our definitions ensure that $w'_{\{i,s\}}$ begins with either

$$i^{k-1}sisis^t i^d \cdots \text{ or } i^{k-1}siisi^d s^t \cdots,$$

where $1 \leq i, s \leq k-1$, i^d and s^t came from the subword w, and the rest of $w'_{\{i,s\}}$ avoids 1^k-matches. Thus, w' represents the edge is. Similarly, our definitions ensure that $w'_{\{j,m\}}$ begins with either $mmjm^t j^d \cdots$ or $mjmj^d m^t \cdots$, where $1 \leq i, s \leq k-1$, j^d and m^t came from the subword w, and the rest of $w'_{\{j,m\}}$ avoids 1^k-matches. Thus, w' represents the edge jm. □

To proceed, we make a convention that if ij denotes an edge in a graph then $i < j$. Also, for a word $w = w_1 w_2 \cdots w_k \in \{1, 2\}^*$ we let $w[i, j]$ denote the word obtained from w by using the substitution: $1 \to i$ and $2 \to j$ for some letters i and j.

Theorem 6.2.3. *([90]) Any graph $G = (V, E)$ is u-representable assuming that $u = 1^a 2^b 1 u_{a+b+2} u_{a+b+3} \cdots u_k$, where $a, b \geq 1$ and $u_{a+b+2} u_{a+b+3} \cdots u_k \in \{1, 2\}^*$.*

Proof. Let $V = \{1, \ldots, n\}$. If $G = K_n$ then by Lemma 6.2.1 G is u-representable. We proceed by induction on the number of edges in a graph with the base case being K_n. Our goal is to show that if G is u-representable, then the graph G' obtained from G by removing any edge ij $(i < j)$ is also u-representable.

Suppose that w u-represents G. We claim that the word

$$w' = u[i, j] 1^{b+1} 2^{b+1} \cdots n^{b+1} w$$

u-represents G'.

Indeed, the vertices i and j are not connected any more because $w'_{\{i,j\}}$ contains u (formed by the k leftmost elements of $w'_{\{i,j\}}$). Also, no new edge can be created in G because w is a subword of w'. Thus, we only need to show that each edge ms $(m < s)$ u-represented by w is still u-represented by w' if $\{m, s\} \neq \{i, j\}$. We have five cases to consider.

1. Suppose $m = i$ and $s \neq j$ $(i < s)$. In this case, $w'_{\{i,s\}} = i^d s^{b+1} w_{\{i,s\}}$, where $d \geq a + b + 2$. Because of s^{b+1}, there is no u-match in $w'_{\{i,s\}}$ that begins to the left of $w_{\{i,s\}}$, and because $w_{\{i,s\}}$ itself has no u-matches, w' u-represents the edge is.

2. Suppose $s = i$ and $m < i$. In this case, $w'_{\{m,i\}} = i^d m^{b+1} i^{b+1} w_{\{m,i\}}$, where $d \geq a + 1$. Because of i^{b+1}, there is no u-match in $w'_{\{m,i\}}$ that begins to the left of $w_{\{m,i\}}$, and because $w_{\{m,i\}}$ itself has no u-matches, w' u-represents the edge mi.

3. Suppose $m \neq i$ and $s = j$ $(m < j)$. This case is essentially the same as (2) after the substitution $i \to j$.

4. Suppose $m = j$ and $s > j$. This case is essentially the same as (1) after the substitution $i \to j$.

5. Suppose $m, s \notin \{i, j\}$. In this case, $w'_{\{m,s\}} = m^{b+1} s^{b+1} w_{\{m,s\}}$. Because of s^{b+1}, there is no u-match in $w'_{\{m,s\}}$ that begins to the left of $w_{\{m,s\}}$, and because $w_{\{m,s\}}$ itself has no u-matches, w' u-represents the edge ms.

We have proved that w' u-represents G', as desired. □

Theorem 6.2.4. *([90]) Let $u = 1^{k-1}2$, where $k \geq 3$. Then every graph $G = (V, E)$ is u-representable.*

Proof. Let $V = \{1, \dots, n\}$. If $G = K_n$ then by Lemma 6.2.1 G is u-representable. We proceed by induction on the number of edges in a graph with the base case being K_n. Our goal is to show that if G is u-representable, then the graph G' obtained from G by removing any edge ij $(i < j)$ is also u-representable.

Suppose that w u-represents G. We claim that the word w' defined as

$$i^{k-2}(i+1)(i+2)\cdots(j-1)(j+1)(j+2)\cdots nij(j+1)\cdots n(i+1)(i+2)\cdots nw$$

u-represents G'.

Indeed, the vertices i and j are not connected any more because $w'_{\{i,j\}}$ contains $i^{k-1}j$. Also, no new edge can be created in G because w is a subword of w'. Thus, we only need to show that each edge ms $(m < s)$ u-represented by w is still u-represented by w' if $\{m, s\} \neq \{i, j\}$. We have five cases to consider.

1. Suppose $m = i$ and $s \neq j$ $(i < s)$. In this case, $w'_{\{i,s\}} = i^{k-2}sissw_{\{i,s\}}$ or $w'_{\{i,s\}} = i^{k-2}sisw_{\{i,s\}}$ depending on whether $s > j$ or not, respectively. Because of the subword sis observed in both cases, there is no u-match in $w'_{\{i,s\}}$ that begins to the left of $w_{\{i,s\}}$, and because $w_{\{i,s\}}$ itself has no u-matches, w' u-represents the edge is.

2. Suppose $s = i$ and $m < i$. However, in this case no such m appears to the left of w in w', and thus there is no u-match in $w'_{\{m,i\}}$ that begins to the left of $w_{\{m,i\}}$. Since $w_{\{m,i\}}$ has no u-match, w' u-represents the edge mi.

3. Suppose $m \neq i$ and $s = j$ $(m < j)$. In this case, we can assume that $m > i$ since otherwise we have a case similar to (2). We have that $w'_{\{m,j\}} = mjmjw_{\{m,j\}}$. Clearly, no u-match can begin at $mjmj$. Since $w_{\{m,j\}}$ has no u-match, w' u-represents the edge mj.

4. Suppose $m = j$ and $s > j$. In this case, we have $w'_{\{j,s\}} = sjsjsw_{\{m,j\}}$. Clearly, no u-match can begin at one of the letters in the subword $sjsjs$. Since $w_{\{j,s\}}$ has no u-match, w' u-represents the edge js.

5. Suppose $m, s \notin \{i, j\}$. In this case, we can assume that $m > i$ since otherwise we have a case similar to (2). We have three subcases to consider here.

   - If $i < m < s < j$ then $w'_{\{m,s\}} = msmsw_{\{m,s\}}$.
   - If $i < m < j < s$ then $w'_{\{m,s\}} = mssmsw_{\{m,s\}}$.
   - If $j < m < s$ then $w'_{\{m,s\}} = msmsmsw_{\{m,s\}}$.

   In either case, it is easy to see that u-match cannot begin to the left of $w_{\{m,s\}}$, and since $w_{\{m,s\}}$ itself has no u-match, w' u-represents the edge ms.

We have proved that w' u-represents G', as desired. □

Theorem 6.2.5. *([90]) Let $u = 1^a 2^b$, where $a, b \geq 2$ and $a + b = k \geq 3$. Then every graph $G = (V, E)$ is u-representable.*

Proof. Let $V = \{1, \dots, n\}$. If $G = K_n$ then by Lemma 6.2.1 G is u-representable. We proceed by induction on the number of edges in a graph with the base case being K_n. Our goal is to show that if G is u-representable, then the graph G' obtained from G by removing any edge ij ($i < j$) is also u-representable.

Suppose that w u-represents G. We claim that the word

$$w' = u[i,j]12 \cdots n12 \cdots nw$$

u-represents G'.

Indeed, the vertices i and j are not connected any more because $w'_{\{i,j\}}$ contains u (formed by the k leftmost elements of $w'_{\{i,j\}}$). Also, no new edge can be created in G because w is a subword of w'. Thus, we only need to show that each edge ms ($m < s$) u-represented by w is still u-represented by w' if $\{m, s\} \neq \{i, j\}$. We have five cases to consider.

1. Suppose $m = i$ and $s \neq j$ ($i < s$). In this case, $w'_{\{i,s\}} = i^{a+1}sisw_{\{i,s\}}$. Because of the subword sis, keeping in mind that $a, b \geq 2$, there is no u-match in $w'_{\{i,s\}}$ that begins to the left of $w_{\{i,s\}}$, and because $w_{\{i,s\}}$ itself has no u-matches, w' u-represents the edge is.

2. Suppose $s = i$ and $m < i$. In this case, $w'_{\{m,i\}} = i^{a+1}mimiw_{\{m,i\}}$. Because $a, b \geq 2$, there is no u-match in $w'_{\{m,i\}}$ that begins to the left of $w_{\{m,i\}}$, and because $w_{\{m,i\}}$ itself has no u-matches, w' u-represents the edge mi.

3. Suppose $m \neq i$ and $s = j$ ($m < j$). This case is essentially the same as (2) after the substitution $i \to j$.

4. Suppose $m = j$ and $s > j$. This case is essentially the same as (1) after the substitution $i \to j$.

5. Suppose $m, s \notin \{i, j\}$. In this case, $w'_{\{m,s\}} = msmsw_{\{m,s\}}$. Because $a, b \geq 2$, there is no u-match in $w'_{\{m,s\}}$ that begins to the left of $w_{\{m,s\}}$, and because $w_{\{m,s\}}$ itself has no u-matches, w' u-represents the edge ms.

We have proved that w' u-represents G', as desired. □

6.3 12-Representable Graphs

In this section, we begin the study of 12-representable graphs. Our first observation is that all graphs on four or fewer vertices are 12-representable. This follows from Theorem 6.5.3 below and the fact that any graph on at most four vertices is a permutation graph. However, in Figure 6.4 we provide 12-representations (by permutations) of small graphs. Note that appropriate labellings have been chosen for the graphs, which is essential in some cases, as will be clear from what follows.

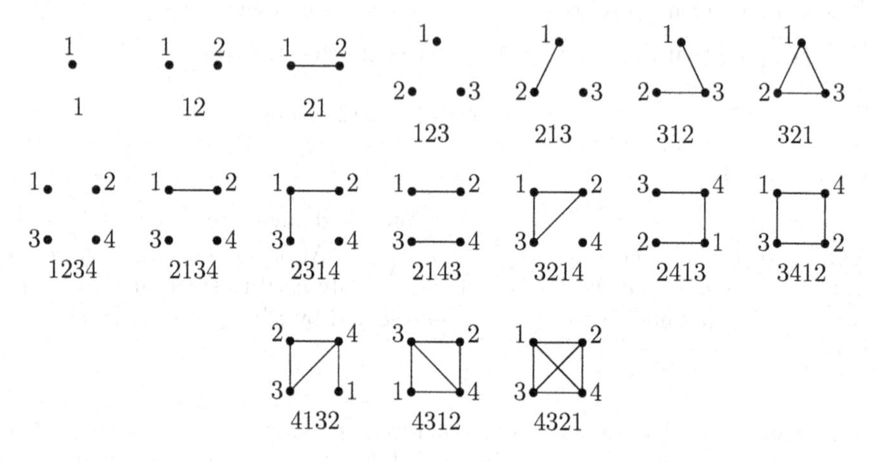

Figure 6.4: Graphs on at most four vertices and their 12-representations

It turns out that any 12-representable graph can be represented by a word having at most two copies of each letter, as is shown by the following theorem.

Theorem 6.3.1. *([84]) If a graph G is 12-representable, then there exists a word 12-representing G in which each letter occurs at most twice.*

Proof. Let a word w 12-represent G. Suppose that a letter j occurs in w at least twice and $w = AjBjC$, where A and C do not contain any copies of j. All non-edges ij with $i < j$ (resp., $i > j$) are necessarily obtained by having a copy of the letter i in AB (resp., BC), and thus all copies of the letter j in B, if any, can be struck out without changing edges/non-edges involving j. □

As a direct corollary to Theorem 6.3.1, we have the following statement.

Corollary 6.3.2. *([84]) If a graph G is 12-representable, then there exists a word 12-representing G in which each letter occurs any given number of times larger than two.*

Proof. Let a word w 12-represent G. By Theorem 6.3.1, we can assume that any letter i occurs at most twice in w. Now, if we replace a copy of i occurring in w by $ii\cdots i$ giving the desired total number of is in w, we will obtain a word that 12-represents G. Alternatively, if i occurs at least twice, we can add new is, if needed, in any positions between the leftmost and rightmost occurrences of i. □

Definition 6.3.3. A vertex x is a *twin* of a vertex y if $N(x) = N(y)$, that is, if the neighbourhoods of x and y coincide.

Example 6.3.4. Figure 6.5 shows adding a twin $1'$ of the vertex 1 in the graph to the left.

Figure 6.5: Adding a twin of a vertex in a graph

Lemma 6.3.5. *([84]) If a graph $G = (V, E)$ is 12-representable and $i \in V \subset \mathbb{P}$, then the graph H obtained by adding a twin of i is also 12-representable.*

Proof. In a labelled copy G' of G realizing 12-representability of G, we let each label $j > i$ become $j + 1$ while keeping all other labels the same. Then we add a twin of i and label it $i + 1$ to obtain a labelled copy H' of H. To see that H' realizes 12-representability of H, we simply substitute each occurrence of i in a word w 12-representing G by $i(i + 1)$, and we replace each letter $j > i$ by $j + 1$, to obtain the word w'. Clearly, from w', ia is an edge in H if and only if $(i + 1)a$ is an edge in H, and i and $i + 1$ are not connected in H. Thus, w' 12-represents H. □

As a direct corollary to Lemma 6.3.5 we have the following remark.

Remark 6.3.6. While studying 12-representability of a graph G, one can assume that there are no two vertices in G that are twins of each other. In particular, if G is a tree, then we can assume that no leaf of G has a sibling.

Next we shall consider labelled graphs I_3, J_4 and Q_4 pictured in Figure 6.6. These graphs will play a key role in determining which graphs are 12-representable.

Theorem 6.3.7. *([84]) Let $G = (V, E)$ be a labelled graph such that $V \subset \mathbb{P}$. Then if G has an induced subgraph H such that* red(H) *is equal to one of I_3, J_4 or Q_4, then G is not 12-representable.*

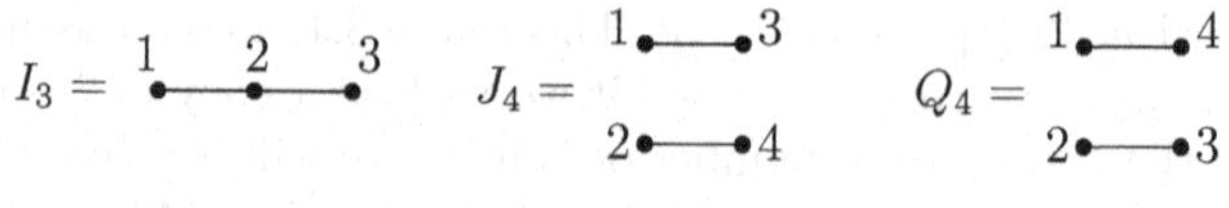

$I_3 =$ 1 — 2 — 3 $J_4 =$ 1 — 3, 2 — 4 $Q_4 =$ 1 — 4, 2 — 3

Figure 6.6: The graphs I_3, J_4 and Q_4

Proof. First, suppose that G has a subgraph H such that $\text{red}(H) = I_3$. Thus H must be of the form $H = (\{i,j,k\},\{ij,jk\})$ where $i < j < k$. Now, for a contradiction, suppose that $w = w_1 \cdots w_n$ 12-represents G. Let w_m be the leftmost occurrence of j in w. Then since $ij \in E$, it must be the case that i does not occur in $w_1 \cdots w_{m-1}$, and since $jk \in E$, it must be the case that k does not occur in $w_{m+1} \cdots w_n$. Let w_s be the rightmost occurrence of k in w. It follows that $s < m$ since $jk \in E$. But since $ik \notin E$, it must be the case that i occurs in $w_1 \cdots w_{s-1}$ which is a contradiction with $ij \in E$.

Next, suppose that G has a subgraph H such that $\text{red}(H) = J_4$. Thus, H must be of the form $H = (\{i,j,k,\ell\},\{ik,j\ell\})$ where $i < j < k < \ell$. Again, for a contradiction, suppose that $w = w_1 \cdots w_n$ 12-represents G. Let w_t be the rightmost occurrence of k in w. Then since $ik \in E$, it must be the case that i does not occur in $w_1 \cdots w_{t-1}$, and since $jk \notin E$, it must be the case that j occurs in $w_1 \cdots w_{t-1}$. Let w_s be the leftmost occurrence of j in w. It follows that $s < t$. But since $j\ell \in E$, it must be the case that ℓ does not occur in $w_{s+1} \cdots w_n$. Next, let w_r be the rightmost occurrence of ℓ in w. Then it must be the case that $r < s$. But since $i\ell \notin E$, it must be the case that i occurs in $w_1 \cdots w_{r-1}$, which is a contradiction with $ik \in E$.

Finally, suppose that G has a subgraph H such that $\text{red}(H) = Q_4$. Thus, H must be of the form $H = (\{i,j,k,\ell\},\{i\ell,jk\})$ where $i < j < k < \ell$. Again, for a contradiction, suppose that $w = w_1 \cdots w_n$ 12-represents G. Let w_t be the rightmost occurrence of ℓ in w. Then, since $i\ell \in E$, it must be the case that i does not occur in $w_1 \cdots w_{t-1}$ and since $j\ell \notin E$, it must be the case that j occurs in $w_1 \cdots w_{t-1}$. Let w_s be the leftmost occurrence of j in w. It follows that $s < t$. But since $jk \in E$, it must be the case that k does not occur in $w_{s+1} \cdots w_n$. Next let w_r be the rightmost occurrence of k in w. Then it must be the case that $r < s$. But since $ik \notin E$, it must be the case that i occurs in $w_1 \cdots w_{r-1}$, which is a contradiction with $i\ell \in E$. □

An immediate corollary to Theorem 6.3.7 is that in a 12-representable labelled graph, the labels must alternate in size through any induced path in the graph.

Definition 6.3.8. A *min-end-path* in a labelled graph $G = (V, E)$ with $V \subset P$ is an induced path in which the two smallest vertices are the endpoints.

The following theorem appeared in the original version of [84].

Theorem 6.3.9. *Let $G = (V, E)$ be a labelled graph where $V \subset \mathbb{P}$. If G has an induced subgraph H which is a min-end-path of length 3 or more, then G is not 12-representable.*

Proof. Suppose that H is of length greater than 3. Then H must be of the form pictured at the top of Figure 6.7 where $n \geq 3$ and $i < j < x_s$ for all $s = 1, \ldots, n$. In such a situation, the subgraph induced by $\{i, j, x_1, x_n\}$ would be reduced to J_4 or Q_4, so that G is not 12-representable by Theorem 6.3.7.

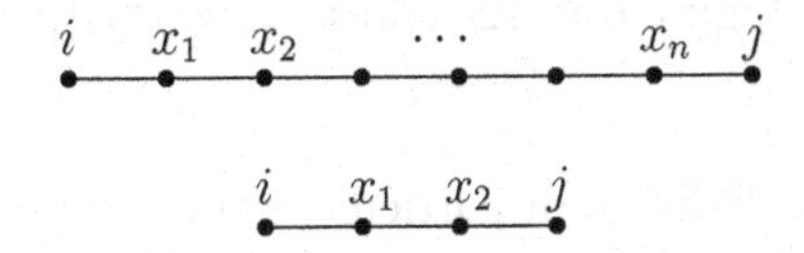

Figure 6.7: Min-end-paths

Suppose that H is of length 3. Then H must be of the form pictured at the bottom of Figure 6.7, where $i < j < x_s$ for $s = 1, 2$. If $x_1 < x_2$, then the subgraph induced by $\{i, x_1, x_2\}$ would reduce to I_3 and if $x_1 > x_2$, then the subgraph induced by $\{x_1, x_2, j\}$ would reduce to I_3. In either case, it follows from Theorem 6.3.7 that G is not 12-representable. □

We note that Theorem 6.3.9 does not say that paths are not 12-representable, but that only certain labelled paths are not 12-representable. In fact, the next result shows that all paths are 12-representable. This result will be generalized in Section 6.4, where we classify all 12-representable trees.

The following theorem appeared in the original version of [84].

Theorem 6.3.10. *All paths are 12-representable.*

Proof. In this case, the proof is best explained through an example. Consider a path of length 9, which will have 10 vertices.

One can imagine that we start with the matching on $\{1, 2, \ldots, 10\}$ which has an edge from $2i - 1$ to $2i$ for $i = 1, \ldots 5$. This is pictured on the left in Figure 6.8. Clearly, this graph is represented by the word $v = 21436587(10)9$. Then we add edges from 1 to 4, 3 to 6, 5 to 8, and 7 to 10 to form a path of length 9. We can then modify v to obtain a word w representing this path by swapping 1 and 4, 3 and 6, 5 and 8, and 7 and 10. Clearly, the word $w = 2416385(10)79$ 12-represents the path. □

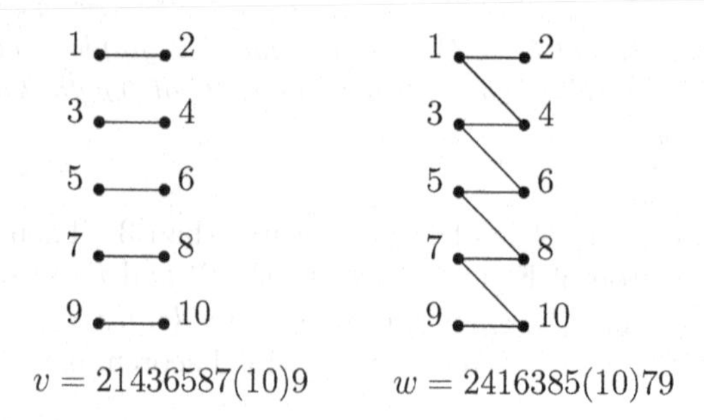

Figure 6.8: 12-representing a path

Remark 6.3.11. ([84]) Paths are a particular type of trees called *caterpillars*; we define the notion of a caterpillar in Section 6.4. It is known that each caterpillar is a permutation graph defined in Definition 2.2.14: e.g. see

`http://www.graphclasses.org/classes/gc_784.html.`

However, all permutation graphs are 12-representable due to Theorem 6.5.2 below. Thus, we have an alternative way to prove Theorem 6.3.10, which would not provide an explicit way to represent paths though.

Theorem 6.3.12. *([84]) Let $G = (V_G, E_G)$ and $H = (V_H, E_H)$, where $V_G, V_H \subset \mathbb{P}$, be 12-representable graphs and G' and H' realize G and H, respectively. Also, let $x \in V_G$ and $y \in V_H$ receive the smallest or the highest label in G' and H', respectively. Then the graph $G \cup H \cup xy = (V_G \cup V_H, E_G \cup E_H \cup \{xy\})$, that is, the graph obtained from non-overlapping copies of G and H by adding the edge xy, is 12-representable.*

Proof. Suppose, without loss of generality, that $V_G = \{1, 2, \ldots, k\}$ and $V_H = \{1, 2, \ldots, \ell\}$. Taking the supplement of graphs G and H, if necessary, by Corollary 6.1.15 we can assume that x is labelled by k and y is labelled by 1 in G' and H', respectively. Also, suppose that words w_G and w_H 12-represent G and H via G' and H', respectively. In particular, $A(w_G) = \{1, 2, \ldots, k\}$ and $A(w_H) = \{1, 2, \ldots, \ell\}$. Further, suppose that w_H^{+k} is obtained from w_H by increasing each letter by k. Finally, let $w_G(k \to k+1)$ be the word obtained from w_G by replacing each occurrence of k by $k+1$, and $w_H^{+k}(k+1 \to k)$ be the word obtained from w_H^{+k} by replacing each occurrence of $k+1$ by k.

We claim that the word $w = w_G(k \to k+1)w_H^{+k}(k+1 \to k)$, obtained by concatenating two words, 12-represents the graph $G \cup H \cup xy$. Indeed, if one labels the graph G as G', graph H as H', then increases each label in H' by k, and then swap the labels k and $k+1$, one obtains a labelled copy of $G \cup H \cup xy$ 12-represented by w (in particular, there is an edge between the vertices labelled by k and $k+1$, namely, x and y). □

Our next theorem tells us how the vertices in two edge-disjoint subgraphs of a given graph G must be related if G is 12-representable graph. Given two subsets of positive integers A and B, we shall write $A < B$ if $x < y$ whenever $x \in A$ and $y \in B$. Thus $A < B$ if every element of A is less than every element of B.

Theorem 6.3.13. *([84]) Suppose that $G = (V, E)$ is a graph, where $V \subset \mathbb{P}$. Let $G_1 = (V_1, E_1)$ and $G_2 = (V_2, E_2)$ be two connected induced subgraphs of G such that $V_1 \cap V_2 = \emptyset$ and there are no edges in E that connect a vertex in V_1 to a vertex in V_2. If G is 12-representable, $|V_1|, |V_2| \geq 2$, and the smallest element of $V_1 \cup V_2$ lies in V_1, then $V_1 < V_2$.*

Proof. Our assumptions ensure that the subgraph of G induced by $V_1 \cup V_2$ is $G_1 \cup G_2 = (V_1 \cup V_2, E_1 \cup E_2)$. Now consider $\text{red}(G_1 \cup G_2) = (\{1, \ldots, n\}, E_1^* \cup E_2^*)$. Assume it is not the case that $V_1 < V_2$. Then there must exist $k, j \geq 1$ such that in the graph $red(G_1 \cup G_2)$, the vertices $\{1, \ldots, k\}$ are in $\text{red}(G_1)$, the vertices $k+1, \ldots, k+j$ are in $\text{red}(G_2)$, and the vertex $k+j+1$ is in $\text{red}(G_1)$. Then let C equal the non-empty set of all vertices in G_1 which are $\geq k+j+1$. Since G_1 is connected, there must be some edge ij in $\text{red}(G_1 \cup G_2)$, where $i \in \{1, \ldots, k\}$ and $j \in C$. Similarly, since G_2 is connected and $|V_2| \geq 2$, there must be some edge $(k+1)\ell$ in $\text{red}(G_2)$. But then the subgraph of $\text{red}(G_1 \cup G_2)$ induced by $\{i, j, k+1, \ell\}$ reduces to either J_4 or Q_4 since $i < k+1 < j, \ell$. Hence G has an induced subgraph which reduces to either J_4 or Q_4 which would contradict our assumption that G is 12-representable graph. Thus it must be the case that $V_1 < V_2$. □

The following theorem provides examples of non-12-representable graphs. Note that in Figure 6.4 we have shown that C_3 and C_4 are 12-representable. It turns out that those are the only cycle graphs which are 12-representable.

Theorem 6.3.14. *([84]) C_n is not 12-representable for any $n \geq 5$.*

Proof. Suppose for a contradiction that $n \geq 5$ and $G_n = (\{1, \ldots, n\}, E)$ is a labelled version of C_n which is 12-representable. Let $1, x_1, x_2, \ldots, x_{n-1}$ be the labels of vertices as we proceed around the cycle in a clockwise manner. Then since no subgraph of G_n can reduce to I_3 by Theorem 6.3.7, it must follow that the sequence $1, x_1, x_2, \ldots, x_{n-1}, 1$ must be an up-down sequence. That is, we must have that $1 < x_1 > x_2 < x_3 > x_4 < \cdots > x_{n-2} < x_{n-1} > 1$. This is clearly impossible if n is odd. Thus we can assume that n is even. But then consider the position of 2 in the sequence $1, x_1, x_2, \ldots, x_{n-1}, 1$. Clearly, 2 cannot be equal to x_1 or x_{n-1}. But this means that one of the two paths that connect 1 to 2 around the cycle would be a min-end-path of length greater than or equal to 3, which is impossible by Theorem 6.3.9. □

6.4 Characterization of 12-Representable Trees

A *caterpillar* or *caterpillar tree* is a tree in which all the vertices are within distance 1 of a central path. In this section, we need the notion of a *double caterpillar* defined as follows.

Definition 6.4.1. A double caterpillar C is a tree in which all the vertices are within distance 2 of a central, possibly one vertex path. Such a path is called the double caterpillar's *spine* and it can be obtained by first removing all leaves from C, and then removing all leaves from the obtained graph.

Example 6.4.2. See Figure 6.9 for an example of a double caterpillar with spine defined by vertices x, y and z.

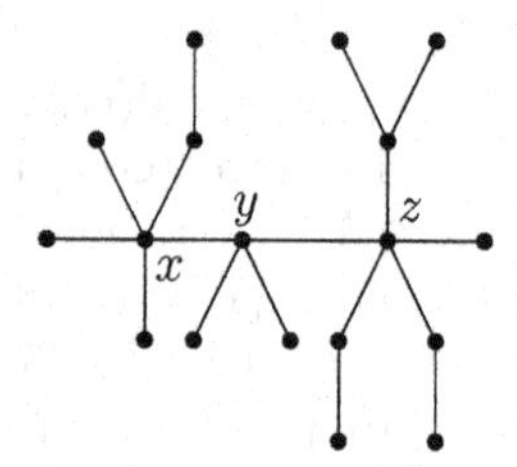

Figure 6.9: A double caterpillar with spine xyz

A *star* or *star tree* S_n is the complete bipartite graph $K_{1,n}$ with $n \geq 0$, where $n = 0$ corresponds to the single-vertex graph K_1. The *centrum* of a star is the all-adjacent vertex in it.

Suppose that a vertex v in a tree T is adjacent to vertices $v_1, v_2, \ldots, v_k$. Removing v together with the k edges incident to it, we obtain a forest $T \backslash v$ whose ith component T_i is induced by the tree containing v_i. We say that the ith component of the forest is *good* if it is a star with centrum at the vertex v_i.

Theorem 6.4.3. *([84]) If a tree T is 12-representable then for any vertex v, at most two components T_i of the forest $T \backslash v$ are not good.*

Proof. By Theorem 6.3.13, without loss of generality we can assume that $T_1 < T_2 < \cdots < T_k$ in a labelled tree T' realizing the 12-representability of T.

Now, suppose that there are three components of the forest $T \backslash v$ that are not good. Without loss of generality, we can assume that these components are T_1, T_2 and T_3. Further, assume that the vertices v, v_1, v_2 and v_3 receive labels r,

$m_1 < m_2 < m_3$, respectively, in T'. Thus, T' contains an induced subgraph F shown schematically in Figure 6.10.

Note that if $m_1 < r < m_3$ then we would obtain a contradiction with Theorem 6.3.7 since $\{v, v_1, v_3\}$ would then induce I_3.

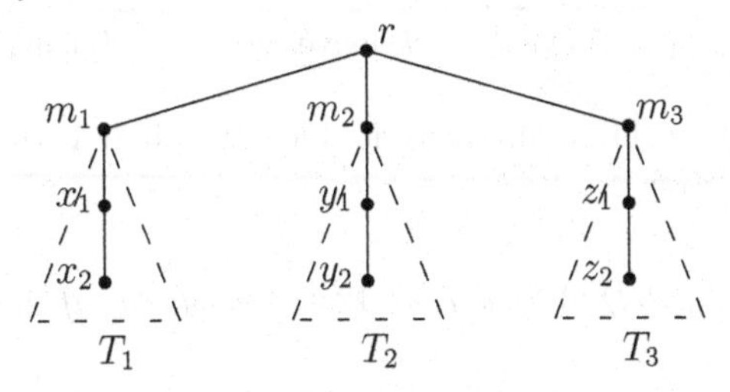

Figure 6.10: The graph F used in the proof of Theorem 6.4.3

We can now assume that $r < m_1$, since in the case $r > m_3$ we can take the supplement of T', apply Lemma 6.1.14, and therefore appear in the situation where $r < m_1$.

Next consider the vertex with label x_1 connected to the vertex labelled m_1 as shown in Figure 6.10. It cannot be that $x_1 > m_1$ because then the subgraph induced by the vertices with labels in $\{m_1, r, x_1\}$ would reduce to I_3. Thus $x_1 < m_1$. Similarly, it cannot be that $x_1 > x_2$ in Figure 6.10 because then the subgraph induced by the vertices with labels in $\{m_1, x_1, x_2\}$ would reduce to I_3. Thus, it must be the case that $r < m_1 > x_1 < x_2$. By the same reasoning we can conclude that $m_2 > y_1 < y_2$ and $m_3 > z_1 < z_2$.

Given our labelling of vertices, we have only three possibilities for r, namely, $r < x_1$, or $x_1 < r < \min\{m_1, x_2\}$, or $\min\{m_1, x_2\} < r < m_1$; if $x_2 > m_1$ then we do not have the last case. We will show that each of these cases leads to a contradiction.

Case 1. $r < x_1$. In this case, the subgraph induced by the vertices with labels in $\{x_1, x_2, r, m_2\}$ is a copy of J_4, which is impossible by Theorem 6.3.7.

Case 2. $x_1 < r < \min\{m_1, x_2\}$. In this case, the subgraph induced by the vertices with labels in $\{x_1, x_2, r, m_2\}$ is a copy of Q_4, which is impossible by Theorem 6.3.7.

Case 3. $\min\{m_1, x_2\} < r < m_1$. In this case, the subgraph induced by the vertices with labels in $\{r, y_1, y_2, m_3\}$ is a copy of Q_4, which is impossible by Theorem 6.3.7. □

We have the following characterization of 12-representable trees.

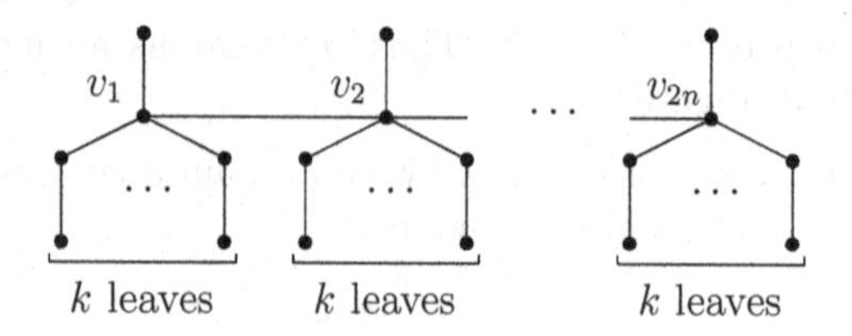

Figure 6.11: A uniform double caterpillar with even spine

Theorem 6.4.4. *([84]) A tree T is 12-representable if and only if it is a double caterpillar.*

Proof. **Necessity.** Suppose that a tree T is 12-representable and T is not a double caterpillar. Further, suppose that $P = v_1v_2 \cdots v_k$ is a path in T of maximum length. Since all trees of diameter 5 are double caterpillars, P has at least six edges, and thus $k \geq 7$. By our assumption, T has a vertex v at distance 3 from P. Suppose that v_i is the closest vertex to v on the path P. Because P is of maximum length, we must have that $i \in \{4, 5, \ldots, k-3\}$. But then in the forest $T \backslash v_i$ at least three components, namely those containing v, v_1 and v_k are not good, and thus by Theorem 6.4.3, T is not 12-representable. Contradiction.

Sufficiency. By Remark 6.3.6, we can assume that no leaf has a sibling. To show that any double caterpillar is 12-representable, we will use induction on the length of the double caterpillar's spine, and will prove the statement for *uniform double caterpillars* $DC(P_{2n})$ with even spines $P_{2n} = v_1v_2 \cdots v_{2n}$ presented schematically in Figure 6.11; the truth of the statement for any other double caterpillar is then obtained by erasing some of the vertices in $DC(P_{2n})$, which will result, using the hereditary property, in a 12-representable graph.

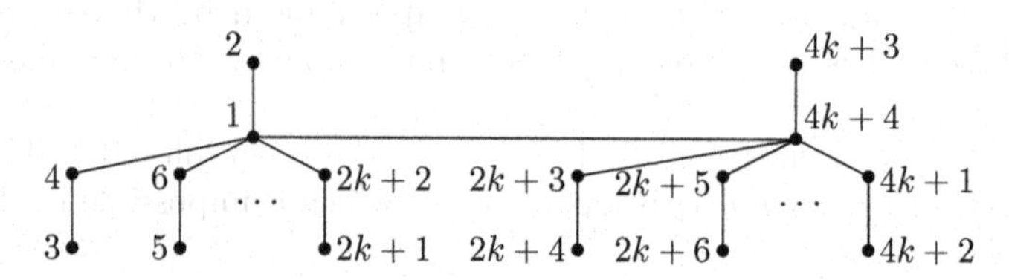

Figure 6.12: The base case for the sufficiency part in the proof of Theorem 6.4.4

We will prove a stronger statement, namely that there is a labelled graph $DC'(P_{2n})$ realizing $DC(P_{2n})$ in which the label of v_1 is 1 and that of v_{2n} is the maximum one. The base case is given by labelling $DC(P_2)$ as in Figure 6.12, and 12-representing this graph by the word

$$24365\cdots(2k+2)(2k+1)(2k+4)(2k+6)\cdots(4k+2)(4k+4)135\cdots$$
$$(2k+1)(2k+4)(2k+3)(2k+6)(2k+5)\cdots(4k+2)(4k+1)(4k+3)$$

stated on two lines. It is straightforward to check that this word has the right alternating properties.

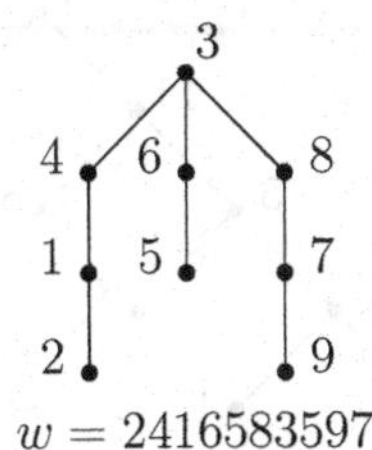

Figure 6.13: A 12-representation of a tree

Now, suppose that we are given a double caterpillar $DC(P_{2n})$. Removing any even edge on $DC(P_{2n})$'s spine, we can break $DC(P_{2n})$ into two double even-spine caterpillars of smaller size $DC(P_{2r})$ and $DC(P_{2(n-r)})$, $1 \leq r \leq n-1$. We can now apply the induction hypothesis to $DC(P_{2r})$ and $DC(P_{2(n-r)})$ and apply the proof of Theorem 6.3.12 to glue these graphs by an edge thus obtaining a realization $DC'(P_{2n})$ of 12-representability of $DC(P_{2n})$. Moreover, the proof of Theorem 6.3.12 ensures that the smallest and largest labels of $DC'(P_{2n})$ will be at the end vertices of the spine P_{2n}. □

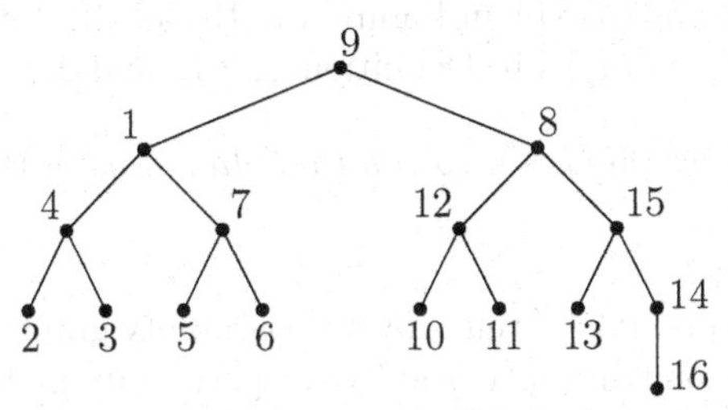

Figure 6.14: A 12-representation of the full binary tree of height 3 plus one vertex

Ignoring the dashed triangles in the graph in Figure 6.10, we obtain the minimum non-12-representable tree on 10 vertices. On the other hand, Figures 6.13, 6.14 and 6.15 give examples of 12-representable trees and words 12-representing them.

Note that in Figure 6.13 the spine is given by the vertices labelled by $3, 4, 8$, in Figure 6.14 by those labelled by $1, 8, 9, 15$, and in Figure 6.15 by the vertices labelled by $5, 7, 11$. Thus, none of the trees in Figures 6.13, 6.14 and 6.15 are labelled in such a way that the spines' end points both receive the extreme values. Also, note that Figure 6.15 provides a step-by-step procedure to 12-represent the *comb graph*. A comb graph is a tree constructed following the pattern in Figure 6.15.

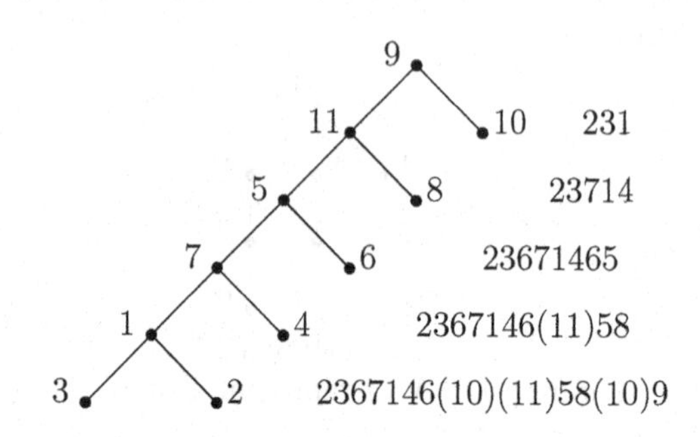

Figure 6.15: 12-representing a comb graph

6.5 Comparing 12-Representable Graphs to Other Classes of Graphs

The goal of this section is to explain Figure 6.1. Recall the notions of comparability graphs and permutation graphs in Definitions 2.2.12 and 2.2.14, respectively.

Definition 6.5.1. A graph G is a *co-comparability graph* if the complement $\overline{G}$ is a comparability graph.

Recall from Theorem 3.4.3 that any comparability graph is word-representable (to be more precise, comparability graphs are permutationally representable), and from Section 3.2, any odd cycle of length 5 or more, despite being a non-comparability graph, is word-representable. Moreover, odd wheels on six or more vertices are non-word-representable by Proposition 3.5.3, and the set of 1^k-representable graphs, for any $k \geq 3$, coincides with the set of all graphs by Theorem 6.2.2. Our next result shows that any 12-representable graph is necessarily a comparability graph.

Theorem 6.5.2. *([84]) If G is a 12-representable graph, then G is a comparability graph.*

Proof. Suppose that G' is a labelled graph realizing 12-representability of G. In particular, by Theorem 6.3.7, any induced path P of length 3 is such that $\text{red}(P) \neq I_3$. We now orient the edges in G' so that if ab is an edge and $a < b$ then the edge is oriented from a to b. This orientation is obviously acyclic. We claim that this orientation is, in fact, transitive, which completes the proof of our theorem. Indeed, if the oriented copy of G' contains a directed path $\vec{P}$ of length 3, say $a \to b \to c$, then we must have the edge $a \to c$ in the graph because otherwise $\text{red}(\vec{P}) = I_3$. □

Recall from Corollary 6.3.2 that a shortest word 12-representing a graph contains at most two copies of each letter. It turns out that those graphs whose 12-representation requires just one copy of each letter are exactly the class of permutation graphs, which is recorded in the next theorem.

Theorem 6.5.3. *([84]) A graph G can be 12-represented by a permutation if and only if G is a permutation graph.*

Proof. The statement follows directly from the definitions. □

Remark 6.5.4. Theorem 6.5.3 shows that the notion of 12-representable graphs is a natural generalization of the notion of permutation graphs: in the former case we deal with multi-permutations to represent graphs, in the latter case with just permutations.

Recall the definition of an interval graph in Definition 2.2.11.

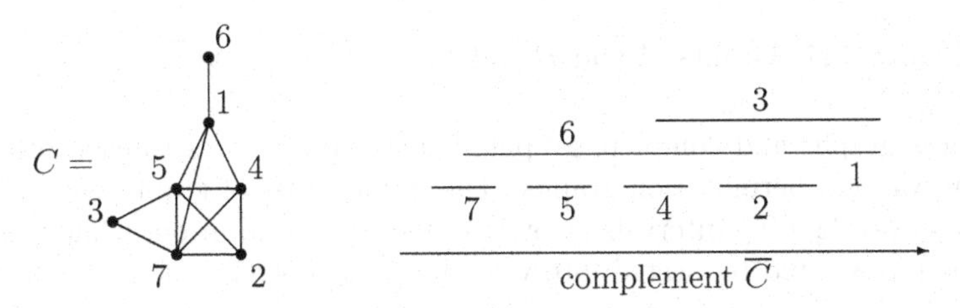

Figure 6.16: The graph C and an interval representation of its complement $\overline{C}$

Theorem 6.5.5. *([84]) If G is a co-interval graph then G is 12-representable.*

Proof. Suppose G is on n vertices. By definition, any interval graph has an interval representation such that the endpoints of the intervals are all distinct. Consider such an interval representation of the complement graph $\overline{G}$. Next, give an interval in this representation the label $n - i + 1$ if the left end of this interval is the ith one from left to right among all endpoints. Such a labelling induces a labelling of

G. We refer to Figure 6.16 for an example of a co-interval graph C and its labelling based on the endpoints of the intervals.

Next, form a word w corresponding to the labelled intervals by going through all interval endpoints (both left and right endpoints) from left to right and recording the labels of the endpoints in the order we meet them. For example, for the labelled interval representation in Figure 6.16, the word w is 7675543426123. Optionally, all occurrences of ii, like 55 in the last word, can be replaced by a single i. We claim that the word w 12-represents G. This follows directly from our labelling of intervals and from the fact that two intervals, I and J, do not overlap if and only if there is an edge between I and J in the graph G. □

To conclude our description of Figure 6.1, we would like to justify that the Venn diagram presented by us is proper, namely that there are strict inclusions of sets and also there is no inclusion of the class of co-interval graphs into the class of permutations graphs, and vice versa, and these classes do overlap. Note that it remains to explain the set inclusions only inside the class of 12-representable graphs since the rest of the diagram has been already explained above.

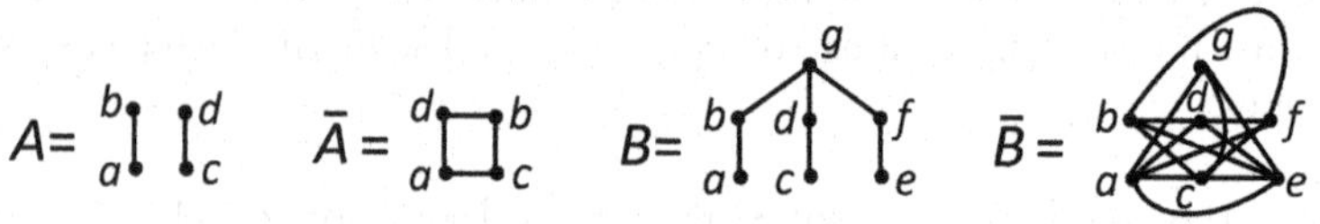

Figure 6.17: Graphs A and B and their complements $\overline{A}$ and $\overline{B}$

Edgeless graphs and complete graphs, being complements to each other, are both co-interval and permutation graphs. The former class of graphs can be represented by non-overlapping intervals or by the identity permutation, while the latter class can be represented by nested intervals and by the reverse of the identity permutation.

In Figure 6.17, there are two graphs, A and B, and their complements $\overline{A}$ and $\overline{B}$. Our goal is to show that the graph A is a permutation graph but not a co-interval graph, while graph B, being 12-representable by Theorem 6.4.4, is neither a co-interval graph nor a permutation graph.

The following theorems are well known results. The proofs of these theorems below are essentially taken from [84].

Theorem 6.5.6. *The graph A in Figure 6.17 is a permutation graph but not a co-interval graph.*

Proof. Letting $a = 1$, $b = 2$, $c = 3$ and $d = 4$ we see that the permutation 2143 represents the graph A, so that it is a permutation graph. The fact that the graph $\overline{A}$ is not an interval graph is well known and is easy to prove. Namely, if $\overline{A}$ were an interval graph, without loss of generality we could assume that the intervals a, b and c are placed as in the left picture in Figure 6.18. But then there is no way to place interval d to make it overlap with a and b but not with c. □

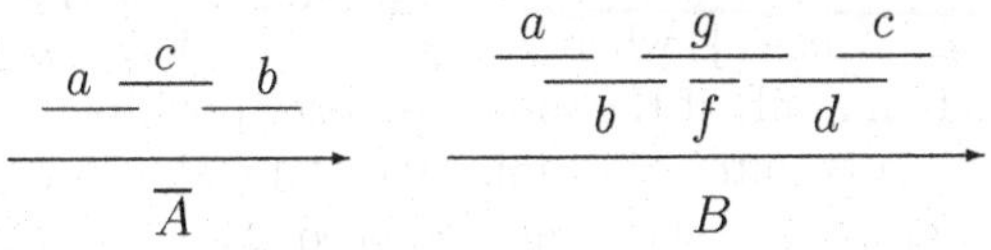

Figure 6.18: Intervals on the real line related to the proofs of Theorems 6.5.6 and 6.5.7

Theorem 6.5.7. *The graph B in Figure 6.17 is neither interval nor co-interval.*

Proof. Let us first consider the graph B. If B were an interval graph, without loss of generality we could assume that the intervals a, b and g are placed as shown in the right-hand picture in Figure 6.18. Another assumption we can make is that the intervals d and c should be placed as shown in the same picture. Then f to be connected with g but disconnected with b and d must be placed between the latter intervals. A contradiction is now obtained with the placing of e which must overlap only with f but not with any other interval, namely, e will be forced to overlap with g as well.

To complete our proof, note that the graph B contains the graph A as an induced subgraph, and thus, by Theorem 6.5.6, B is not co-interval. □

To prove our next theorem, we need the following definition. Recall from Definition 2.2.7 that a graph is chordal if each of its cycles of four or more vertices has a chord, which is an edge that is not part of the cycle but connects two vertices of the cycle. Equivalently, every induced cycle in the graph should have at most three vertices. Also, recall from Theorem 2.2.17 that a graph G is a permutation graph if and only of both G and its complement $\overline{G}$ are comparability graphs.

Again, the following theorem is a well known result, whose proof below is taken from [84].

Theorem 6.5.8. *The graph B in Figure 6.17 is not a permutation graph.*

Proof. Since B is a tree, B is chordal, but it is not an interval graph by Theorem 6.5.7. Since any chordal co-comparability graph is an interval graph [66], we obtain that the complement $\overline{B}$ is not a comparability graph, and thus B is not a permutation graph by Theorem 2.2.17. □

Finally, it remains to provide an example of a co-interval graph which is not a permutation graph. The vertices in our graph G_n are defined by *all* intervals of non-zero length with left endpoints in the set $\{0, 1, \ldots, n\}$ and right endpoints in the set $\{1-\epsilon, 2-\epsilon, \ldots, n-\epsilon\}$, where $\epsilon \in (0, 1)$. Further, two vertices are connected in G_n by an edge if and only if the intervals corresponding to them do *not* overlap. By definition, G_n is a co-interval graph. We claim that for large enough n, G_n is not a permutation graph. To show this, we need the following lemma.

Lemma 6.5.9. *([15]) A graph G is a permutation graph if and only if it is the comparability graph of a partially ordered set that has order dimension at most* 2.

Let us consider a partial order P on the set of all vertices in G_n, where $I < J$ if and only if interval I is entirely to the left of interval J. P is called the *interval order*, and it provides a transitive orientation of G_n. It is shown in Example 8.1.4 in [124] that as n increases, the dimension of P will get larger and larger. Thus, by Lemma 6.5.9, for large enough n, G_n will be a non-permutation graph.

6.6 12-Representation and Grid Graphs

In this section, again based on [84], we consider 12-representation of induced subgraphs of a *grid graph*. Examples of a grid graph and some of its possible induced subgraphs are given in Figure 6.19, where the notions of "*corner graphs*" and "*skew ladder graphs*" were introduced by the authors of [84].

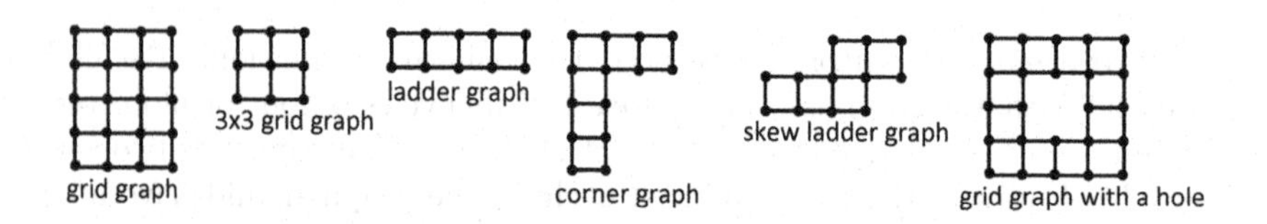

Figure 6.19: Induced subgraphs of a grid graph

Clearly, grid graphs with holes, such as the rightmost graph in Figure 6.19 are not 12-representable because they have large induced cycles, which are not possible in 12-representable graphs by Theorem 6.3.14. In fact, the 3×3 grid graph $G_{3\times3}$ is

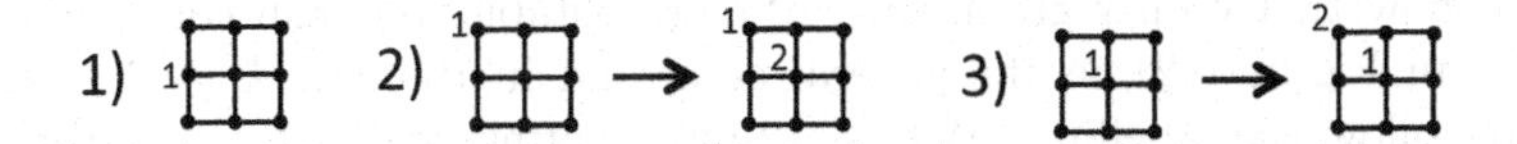

Figure 6.20: No labelled graph can realize the 3×3 grid graph

not 12-representable either. To see this, we can show that no labelling of the graph will result in a labelled graph realizing $G_{3\times3}$.

Indeed, suppose a proper labelling exists. Because of the symmetries, the label 1 is in one of the three positions shown in Figure 6.20. In case 1), to avoid paths of length 3 that can be reduced to I_3, label 2 cannot stand next to label 1, and clearly no matter where label 2 is then placed, we will obtain a copy of J_4 or Q_4 in $G_{3\times3}$ involving labels 1 and 2, and leading to non-12-representability by Theorem 6.3.7. In case 2), label 2 cannot stand next to label 1 to avoid I_3, and moreover, to avoid J_4 or Q_4 involving labels 1 and 2, label 2 must be in the middle. Further, one can see that to avoid I_3, one cannot place label 3 next to labels 1 and 2, and no matter where label 3 will then be placed (in one of the three corner points), there will be a copy of J_4 or Q_4 involving labels 1 and 3, and Theorem 6.3.7 can be applied. Finally, case 3) is similar to case 2) once the observation is made that label 2 must be in one of the corner points to avoid I_3, except now J_4 or Q_4 will be forced to appear using labels 2 and 3.

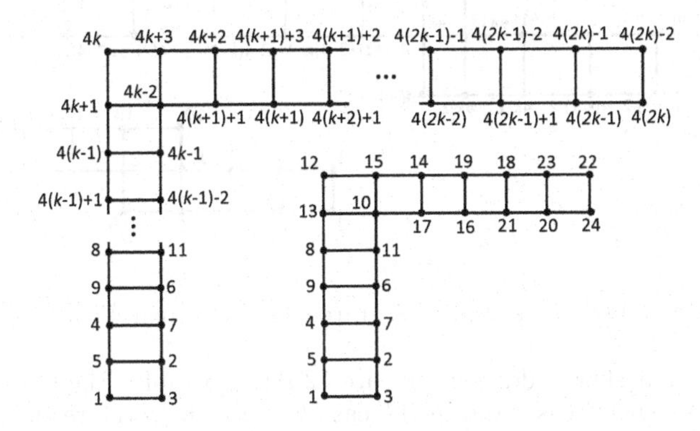

Figure 6.21: Labelling corner graphs to show their 12-representability

The situation with ladder graphs, corner graphs and skew ladder graphs is different. These graphs turn out to be 12-representable. Note that representability for ladder graphs follows from representability of either of the other two classes of graphs.

To show that corner graphs are 12-representable, one can consider labelling as shown in Figure 6.21 in the general case, and in the case $k = 3$ to help the reader to follow the labelling. Words, 12-representing the general (on three lines) and particular cases, respectively, are as follows

$$3.51.72.94.(11)6.\cdots.(4k+1)(4(k-1)).(4k+3)(4k).(4(k+1)+1)(4k-2).\mathbf{4k}.$$

$$(4(k+1)+3)(4k+2).(4(k+2)+1)(4(k+1)).\cdots.$$

$$(4(2k)-1)(4(k+1)+2).(4(2k))(4(2k-1)).(4(2k)-2)$$

and

$$3.51.72.94.(11)6.(13)8.(15)(12).(17)(10).(\mathbf{12}).(19)(14).(21)(16)(23)(18).(24)(20).(22),$$

where the dots are meant to make clearer the pattern in our construction, and the first word is on two lines. Note the corner element in bold that is repeated in our construction. We do not provide a careful justification of why these words work, which can be seen by direct inspection.

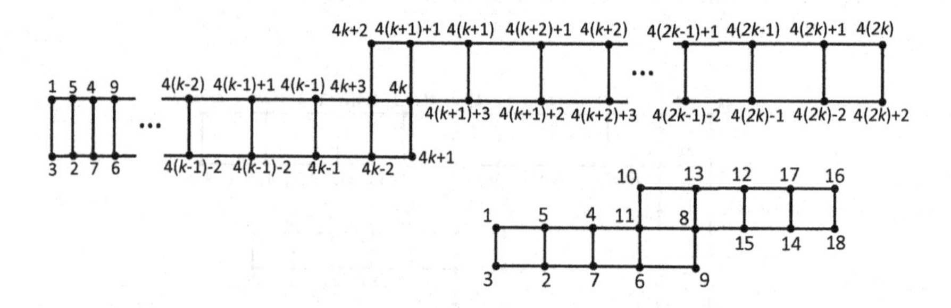

Figure 6.22: Labelling skew ladder graphs to show their 12-representability

To show that skew ladder graphs are 12-representable, Figure 6.22 shows the labelling in the general case, and in the case $k = 2$ to help the reader to follow the labelling. Words 12-representing the general (on three lines) and particular cases, respectively, are as follows

$$3.51.72.94.(11)6.\cdots.(4(k-1)+1)(4(k-2)).(4k-1)(4(k-1)-2).(\mathbf{4k+1}).(4k+3)(4(k-1)).$$

$$(4k+1)(4k-2).(4(k+1)+1)(4k+2).(4(k+1)+3)(4k).(\mathbf{4k+2}).(4(k+2)+1)(4(k+1)).$$

$$(4(k+2)+3)(4(k+1)+2).\cdots.(4(2k)+1)(4(2k-1)).(4(2k)+2)(4(2k-1)+2).(4(2k))$$

and

$$3.51.72.\mathbf{9}.(11)4.96.(13)(10).(15)8.\mathbf{10}.(17)(12).(18)(14).(16),$$

where the first word is on three lines, and again, we indicate the corner elements in bold.

It would be interesting to know whether or not induced subgraphs of a grid graph have a nice 12-representation classification. We leave this as an open problem along with the larger problem of finding a classification of 12-representable graphs.

6.7 Conclusion

The class of word-representable graphs considered in Chapters 3–5 is probably the most interesting part of a much more general theory of graphs representable via pattern avoiding words. In the language of that theory, our word-representable graphs are nothing else but 11-representable graphs, while the general case deals with u-representation of graphs. It is shown that any graph is u-representable assuming that u is of length 3 or more.

The main focus of the chapter is to study so-called 12-representable graphs. A crucial difference between these graphs and word-representable graphs is that labelling of a graph plays an important role, since different labellings of the same graph may result in us being able or unable to find a word-representant. We see that the class of 12-representable graphs is included in the class of comparability graphs, and it includes co-interval and permutation graphs. A nice property of 12-representable graphs is that representation of such a graph requires at most two copies of each letter.

While all trees can easily be word-represented using two copies of each letter, not many trees are 12-representable. It turns out that a tree is 12-representable if and only if it is a double caterpillar.

In Section 7.7, we will introduce a number of other ways to define representability of (hyper)graphs via (patterns in) words and give the notions of various so-called Wilf-equivalencies, which are yet to be studied.

6.8 Exercises

1. The right graph in Figure 6.5 is 12-representable due to Figure 6.4 and Lemma 6.3.5. Relabel the graph in a suitable way to find a word 12-representing it.

2. Using Theorem 6.4.4, explain why the left tree in Figure 6.23 is 12-representable, while the right tree is not.

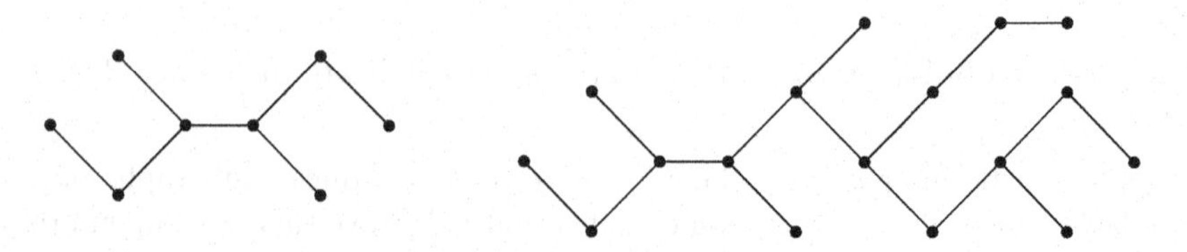

Figure 6.23: Two trees related to Exercise 2

3. Based on Theorem 6.3.7, explain why the labellings of the graphs in Figure 6.24 cannot be used to 12-represent these graphs.

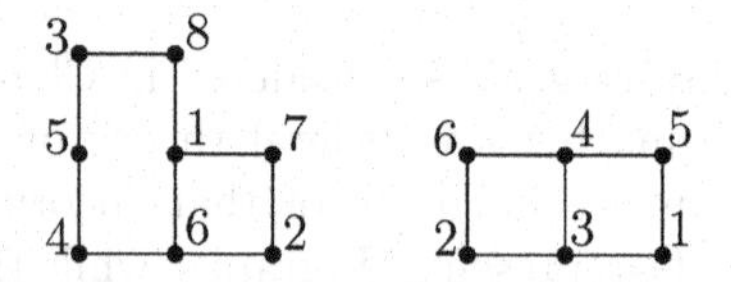

Figure 6.24: Two labelled graphs related to Exercise 3

4. Using the construction to 12-represent the tree in Figure 6.12, 12-represent the left tree in Figure 6.23.

5. Using the construction in Theorem 6.4.4, 12-represent the tree in Figure 6.9.

6. Using the 12-representation of the corner graph on 24 vertices in Figure 6.21, 12-represent the left graph in Figure 6.25.

7. Using the 12-representation of the skew ladder graph on 18 vertices in Figure 6.22, 12-represent the right graph in Figure 6.25.

8. Find a 12-representation of the graph in Figure 6.26.

9. Give your own examples, different from those provided in this chapter (in particular, in Figure 6.1), of a 12-representable graph that is

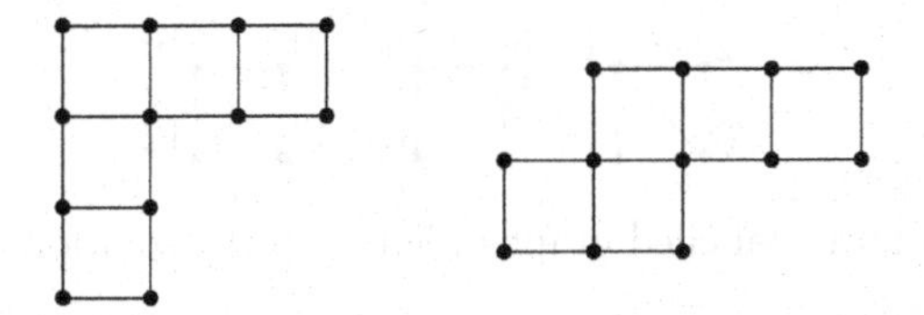

Figure 6.25: A corner graph and a skew ladder graph

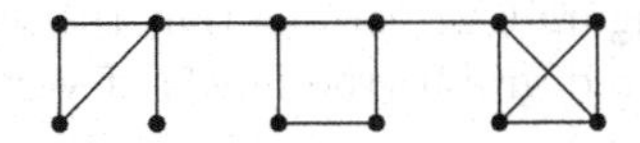

Figure 6.26: The graph related to Exercise 8

- a co-interval graph and a permutation graph,
- neither a co-interval graph nor a permutation graph,
- a permutation graph, but not a co-interval graph,
- a co-interval graph, but not a permutation graph.

6.9 Solutions to Selected Exercises

3. In the left graph, for example, the subgraph induced by the vertices $2, 3, 5, 7$ forms Q_4, while in the right graph, the subgraph induced by the vertices $2, 3, 4$ or $3, 4, 5$ forms I_3.

6. In the word 12-representing the corner graph G on 24 vertices in Figure 6.6 (see Section 6.6) we remove the letters 1–5, 7, 18, 20–24, corresponding to removing the respective vertices from G resulting in a graph G', to obtain the word:
$$w = 9(11)6(13)8(15)(12)(17)(10)(12)(19)(14)(16).$$
The desired 12-representation (of $\text{red}(G')$) is obtained by taking the reduced form of w:
$$\text{red}(w) = 3517296(11)46(12)8(10).$$

8. Consider the three labelled graphs in Figure 6.27, which can be 12-represented by the words 4132, 7856 and (12)(11)(10)9, respectively, following the representations in Figure 6.4. Using the construction of the proof of Theorem 6.3.12, the word 51327846 12-represents the graph G obtained from the leftmost two

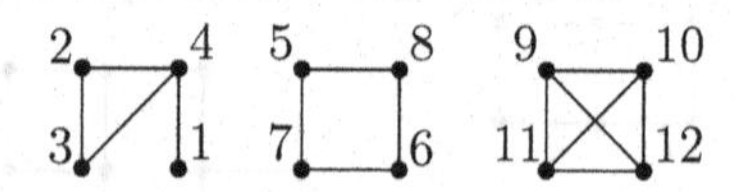

Figure 6.27: Three labelled graphs related to the solution of Exercise 8

graphs by adding the edge 45 and swapping the labels of the vertices 4 and 5. Applying the construction in the proof of Theorem 6.3.12 to G and to the rightmost graph in Figure 6.27, we see that the word 51327946(12)(11)(10)8 12-represents the desired graph (the labels of vertices 8 and 9 are switched, and the edge 89 is added).

Chapter 7

Open Problems and Further Research Directions

In this chapter, we sketch a few directions of research on word-representable graphs and discuss possible approaches to handle some of the problems (see Section 7.6).

In our opinion, one of the most challenging and interesting problems that may lead to potential applications is the following question.

Problem 7.0.1. Suppose that a graph G is word-representable (and a word representing G is given). When can this information about G help us to solve some problems on G that are hard to solve otherwise?

Recall that Theorem 3.4.14 shows that the Maximum Clique problem is polynomially solvable on word-representable graphs, while in general this problem is NP-complete. On the other hand, many classical optimization problems are NP-complete on word-representable graphs (see Table 4.1). What we would like to see with respect to Problem 7.0.1 is the situation when a word representing a graph suggests a better solution to a difficult problem than the graph itself. Witnessing such a situation would give extra motivation to develop further the theory of word-representable graphs.

7.1 A Few Research Directions Without Common Thread

In this section, we list several research directions/open questions without common thread; more open problems and research directions will be offered in the upcoming sections in this chapter.

7.1.1 Characterization by Forbidden Subgraphs

While dealing with a class of graphs, it is a natural fundamental question whether the class can be characterized in terms of obstructions (forbidden elements), unless the class is defined in these terms. One of the most famous such characterizations is *Kuratowski's theorem*, by Polish mathematician Kazimierz Kuratowski, stating that a finite graph is planar if and only if it does not contain the complete graph K_5 or complete bipartite graph $K_{3,3}$ as a minor; see [102]. In Chapter 2, we have seen characterizations of various other classes in terms of other forbidden elements, such as forbidden vertex minors (circle graphs), forbidden induced minors (interval graphs) and forbidden induced subgraphs (line graphs and perfect graphs). For the class of word-representable graphs, no such characterization is known. On the other hand, since this class is hereditary, there must exist a characterization of word-representable graphs in terms of minimal forbidden induced subgraphs. Finding this characterization is a problem of fundamental interest.

Problem 7.1.1. Find the characterization of word-representable graphs in terms of forbidden induced subgraphs.

7.1.2 Word-Representants, 3-Word-representable Graphs and Enumeration

The following open problem is related to Subsection 4.2.1.

Problem 7.1.2. What is the maximum representation number of a graph on n vertices? In other words, how many copies of each letter does a "hardest" graph on a fixed number of vertices require to be word-represented? We know that this number is between $n/2$ and $2n$ for a graph on n vertices (see Subsection 4.2.1).

Section 5.1 offers characterizations of 1- and 2-word-representable graphs. In Section 5.2, we discuss some known results on 3-word-representable graphs. However, a complete characterization of 3-word-representable graphs remains an open problem.

Problem 7.1.3. Characterize 3-word-representable graphs and, more generally, k-word-representable graphs.

As for now, we only know asymptotic enumerations of the word-representable graphs given in Section 5.3. Exact enumerations are unknown.

Problem 7.1.4. How many (k-)word-representable graphs are there?

7.1.3 Squares and Cubes in Word-Representants

Definition 7.1.5. A *square* (resp., *cube*) in a word is two (resp., three) consecutive equal factors. A square (resp., cube) of the form xx (resp., xxx), where x is a letter, is called *trivial.*

Example 7.1.6. The word 2441351352 has two squares, namely, 44 and 135135, while the word 1212124333232322 has three cubes, namely, 121212, 333 and 323232. Note that 44 and 333 are a trivial square and a trivial cube, respectively.

Square-free words have attracted much attention in the literature of combinatorics on words since the work by Axel Thue [131] at the beginning of the twentieth century. This work actually opened the area of combinatorics on words, and established the existence of infinite square-free words over three-letter alphabets. On the other hand, the *Peano word* [91] is an example of a word that contains no cubes except for trivial ones (this fact is easy to prove by mathematical induction).

Consider the edgeless graph O_2 on two vertices. It is not difficult to see that word-representation of such a graph will require the presence of a trivial square, either xx or yy. Thus, there is no *square-free representation* of this graph. Theorem 7.1.8 below shows that O_2 is the only graph any of whose representations requires the presence of trivial squares.

Lemma 7.1.7. *If a word-representable graph G contains at least one edge then its isolated vertices, if any, can be represented without involving trivial squares.*

Proof. For isolated vertices $x, y, z, \ldots$ consider a word-representant $\ldots zyxWxyz \ldots$, where the non-empty word W represents the rest of the graph. Trivial squares are then avoided in representation of the isolated vertices $x, y, z, \ldots$. □

Theorem 7.1.8. *Trivial squares can be avoided in a word-representation of all but the graph O_2.*

Proof. Trivial squares are unavoidable in a word-representation of O_2 by our discussion preceding Lemma 7.1.7.

If a vertex x in a word-representable graph G is incident to an edge, then no representation of G can have xx as a factor. On the other hand, if x is isolated and there is at least one edge in G, then by Lemma 7.1.7 we can represent G without involving xx. Thus, we only need to properly represent an edgeless graph on n vertices, where either $n = 1$ or $n > 2$. In the former case, there is a one-letter word-representant to represent the one-vertex graph, which avoids trivial squares; in the latter case, a trivial square-free representation is given by the word

$$2134 \cdots (n-1)12 \cdots (n-1)n(n-1) \cdots 21.$$

The theorem is proved. □

Our next theorem shows that every graph admits a *cube-free* representation.

Theorem 7.1.9. *For any word-representable graph G, there exists a cube-free word representing G.*

Proof. Suppose that a word $w = uXXXv$, where u, v and X are some words, represents a graph G. Then the word $w' = uXXv$ also represents G. Indeed, for any pair of letters, say (x, y), its (non-)alternation is not changed. To see this, one can consider the following cases:

- $x, y \notin X$. Clearly, the alternation of x and y is the same in w and w'.
- $x \in X$ and $y \notin X$. In this case, x and y alternate neither in w nor in w', because of the presence of the factor XX in both words.
- $x, y \in X$. Recall that $X_{\{x,y\}}$ denotes the word obtained from X by removing all letters except for x and y. Assume, without loss of generality, that the first letter of $X_{\{x,y\}}$ is x. If the last letter of $X_{\{x,y\}}$ is also x, then x and y alternate neither in w nor in w'. Suppose now that the last letter of $X_{\{x,y\}}$ is y. If x and y alternate in w, then clearly they alternate in w'. On the other hand, if x and y do not alternate in w, then the removal of one X from XXX will not make these letters alternate.

The cases above cover all possible situations, and thus both w and w' represent G. Turning all cubes into squares, one by one, will give us the desired cube-free word representing G. □

Problem 7.1.10. Is it true that any word-representable graph can be represented by a word containing no non-trivial squares? If not, describe the class of graphs requiring (non-trivial) squares in all of its representations.

Let us modify slightly, in a natural way, the definition of word-representability as follows: if a graph G is word-representable and x is an isolated vertex in it (x is not connected to any other vertex in G) then x does not appear in a word representing G. With the modified definition in mind, Problem 7.1.10 asks whether we can always find a square-free representation of a word-representable graph.

7.1.4 Word-Representable Graphs and Path-Schemes

There are several classes of graphs that we know to be word-representable, in particular outerplanar, 3-colourable, bipartite and comparability graphs; see Figure 1.1, while, e.g. not all planar graphs are word-representable. Thus, the following research direction can be stated.

Problem 7.1.11. Find (new) examples of (known) classes of graphs which belong to the class of word-representable graphs.

A particular class of interest here is the class of *path-schemes* defined in Section 9.4 below. Path-schemes with one parameter are easily seen to be bipartite graphs, and thus they are comparability graphs, and therefore they are permutationally representable. Moreover, it is not difficult to show by induction on the number of vertices that if a path-scheme has two parameters, then it is 3-colourable. Indeed, adding the new vertex to the left, it is only connected to two already existing vertices and thus there is a free colour we can use for the new vertex. Therefore, by Theorem 4.3.3, any path-scheme with two parameters is word-representable. No word-representability classification is known for path-schemes with three or more parameters.

Problem 7.1.12. Classify word-representable path-schemes with three or more parameters. Give an example of a non-word-representable path-scheme with the fewest parameters.

7.1.5 k-Word-representability and Eulerian Cycles in Regular Graphs

Sergey Avgustinovich suggested the following direction of research in 2011.

Suppose that a word $w = w_1w_2\cdots w_{kn}$ k-represents a graph G on n vertices. In particular, by Definition 3.2.3, each letter occurs in w exactly k times. Then w corresponds to an Eulerian cycle in a $(2k)$-regular multigraph F on n vertices (parallel edges and loops are allowed). Indeed, one can build F as follows.

The set of vertices of F coincides with that of G. The edges of F are defined by the Eulerian cycle corresponding to w. Namely, start with the vertex w_1 in F and move to the vertex w_2 in F thus creating the edge, possibly a loop, w_1w_2. Next move to the vertex w_3 thus creating the edge, possibly a loop, w_2w_3. And so on. Once the vertex w_n is reached, go to the vertex w_1 thus completing the cycle. Therefore, reading w from left to right not only creates F, but also defines an Eulerian cycle in it. For example, see the two words in Figure 7.1, namely *abacdbdc* and *abdcabdc*, which correspond to different Eulerian cycles in the same graph F to the left.

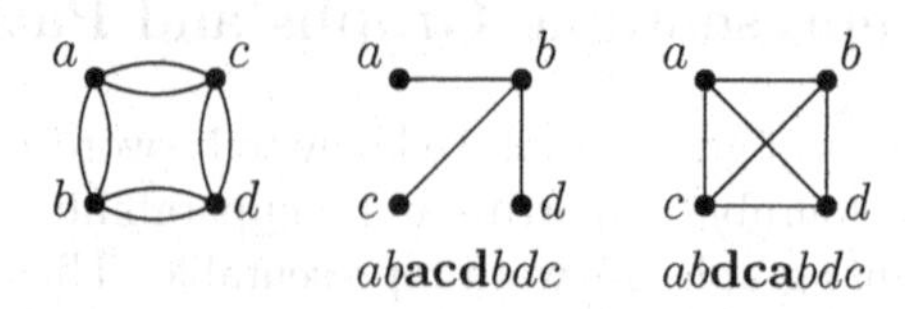

Figure 7.1: A 4-regular graph F (the leftmost graph) and the graphs 2-represented by two Eulerian cycles in F

Let xXx be a factor (see Definition 3.2.8) in w such that x does not occur in X. Reversing the order of letters in xXx, that is, considering $r(xXx) = xr(X)x$, corresponds to changing the orientation of a cycle beginning and ending at x in F. It was proved by Abrham and Kotzig [1] that using the operation of changing the orientation of a directed cycle, one can obtain any Eulerian cycle of F from any of its other Eulerian cycles. Of course, reversing a factor in w will normally results in a word k-representing a different graph than G. For example, if G is the middle graph in Figure 7.1 represented by the word *abacdbdc*, then reversing the factor between two *b*s results in the word *abdcabdc* representing the complete graph K_4. A natural problem to state is then as follows.

Problem 7.1.13. Describe the class of graphs word-represented by words corresponding to all Eulerian cycles in a $(2k)$-regular graph. An alternative formulation is: given a graph G that is k-represented by a word w, describe the class of graphs represented by all the words obtained from w by reversing factors of the form xXx, where x does not occur in X.

7.1.6 Regular Non-word-representable Graphs

Neither the non-word-representable graph W_5 nor any of the non-word-representable graphs in Figure 3.9 are regular. Examples of regular non-word-representable graphs on eight vertices were found by Herman Z.Q. Chen [34] and they are presented in Figure 7.2. This suggests the following direction of research.

Problem 7.1.14. Classify regular non-word-representable graphs.

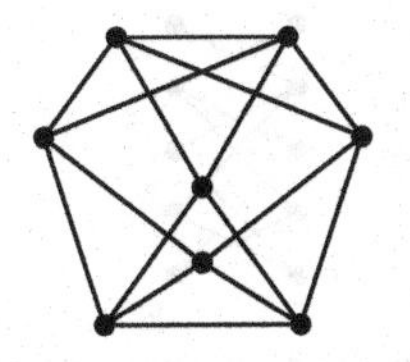
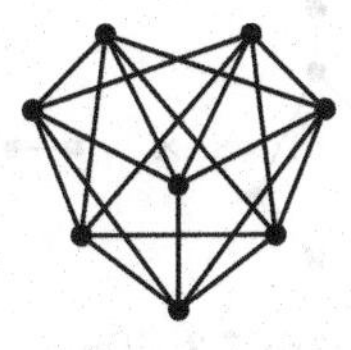

Figure 7.2: Regular non-word-representable graphs

7.2 More Directions Without Common Thread

7.2.1 Planar Graphs

It follows, for example, from Proposition 3.5.3 that not all planar graphs are word-representable. On the other hand, by Theorem 4.3.9 triangle-free planar graphs are word-representable. The following remains a challenging open problem.

Problem 7.2.1. Characterize all word-representable planar graphs.

Note that in odd wheels on six or more vertices (that are non-word-representable by Proposition 3.5.3) and in the non-word-representable planar graphs in Figure 3.9, there is always a pair of triangles that share an edge. As a step towards solving Problem 7.2.1 one could consider the following problem suggested by Xing Peng.

Problem 7.2.2. Do there exist non-word-representable planar graphs in which no two triangles share an edge? If the answer to this question is "yes", then characterize all such graphs.

7.2.2 Tensor Product of Two Graphs

Definition 7.2.3. For two graphs, $G = (V(G), E(G))$ and $H = (V(H), E(H))$, their *tensor product* $G \times H$ is a graph such that

- the vertex set of $G \times H$ is the Cartesian product $V(G) \times V(H)$; and
- two vertices (u, u') and (v, v') are adjacent in $G \times H$ if and only if u' is adjacent to v' and u is adjacent to v.

The tensor product is also called the *direct product, categorical product, cardinal product, relational product, Kronecker product, weak direct product* or *conjunction.*

Example 7.2.4. See Figure 7.3 for an example of the tensor product of two graphs.

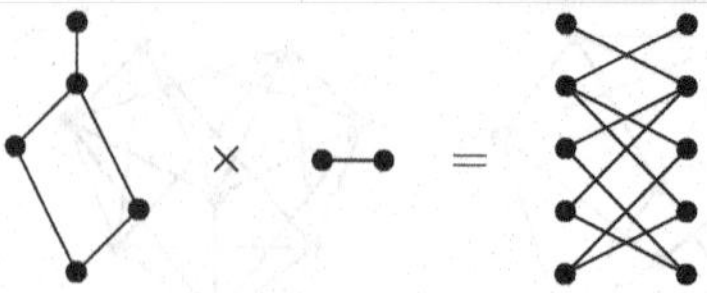

Figure 7.3: Tensor product of two graphs

Theorem 5.4.10 in Subsection 5.4.5 states that the Cartesian product of two word-representable graphs is always word-representable, while clearly if one graph is non-word-representable then the Cartesian product is non-word-representable as well. The same applies to the rooted product of two graphs by Theorem 5.4.13. In the case of the tensor product, we do not know answers to similar questions.

Problem 7.2.5. Can anything definite be said about taking the tensor product of two word-representable graphs? Or the tensor product of a word-representable and a non-word-representable graph? Or the tensor product of two non-word-representable graphs?

The same questions as those in Problem 7.2.5 can be asked about other notions of graph products, e.g. *lexicographical product*, not to be defined in this book.

7.2.3 Circle Graphs

Recall that circle graphs, defined in Subsection 2.2.4, are 2-representable by Theorem 5.1.7. Answering the question in the following problem will likely involve geometric considerations about placing chords corresponding to graph vertices on a circle, and either providing a counterexample for the negative answer, or some sort of geometric argument for the positive answer.

Problem 7.2.6. Given a circle graph G with an edge xy. Is it always possible to find a 2-representation of G in which x and y stand next to each other?

If the question in Problem 7.2.6 were answered in the affirmative, then we could prove that a 3-subdivision (see Definition 5.2.4) of any edge in a circle graph results in a circle graph. This would be done via replacing the factor xy in a word representing G (such a factor would exist) by the factor $uyvuxv$, leading to removing the edge xy from G and inserting a path of length 3, $xvuy$, between x and y. The same substitution was used in proving Theorem 5.2.21 about ladder graphs.

7.2.4 Classes $\mathcal{R}_1$, $\mathcal{R}_2$ and $\mathcal{R}_3$

Subsection 5.2.4 does not answer the following question.

Problem 7.2.7. Can each bipartite graph be 3-word-represented? Namely, does each bipartite graph belong to the union of the sets $\mathcal{R}_1 \cup \mathcal{R}_2 \cup \mathcal{R}_3$?

We suspect the answer to the question in Problem 7.2.7 to be negative. A good candidate for a counterexample should be the crown graph $H_{k,k}$ for $k \geq 5$; see Section 7.4 and, in particular, Problem 7.4.2.

7.2.5 Representation Number and Colourability

The discussion in Subsection 5.2.4 shows that 3-word-representable graphs are not c-colourable for a constant c, while the graph G_k in Definition 4.2.4 having representation number $\lfloor \frac{n}{2} \rfloor$ (by Theorem 4.2.6) is 3-colourable. This observation led us to the following questions.

Problem 7.2.8. Are there constants k and c such that graphs with representation number k are necessarily c-colourable? Also, do there exist constants k and c, $0 < k \leq 1$, such that graphs with representation number kn are necessarily c-colourable? (Recall that by Theorem 4.2.1 each word-representable graph requires at most n copies of each letter to be represented.)

7.2.6 Preference Orders

Even though our main concern so far has been in the *existence* of a word representing a given graph, we note that typically, there are several non-equivalent words representing the same graph. Thus, one can state the following research direction.

Problem 7.2.9. Given a word-representable graph G, study *(preference) orders* on its word-representants. In particular, what is the lexicographically smallest word-representant?

A motivation to work on Problem 7.2.9 comes, e.g. from *robot plan execution* related to scheduling problems mentioned in Section 1.6. For example, implementing a *security patrol plan*, again mentioned in Section 1.6, corresponds to reading a word-representant from left to right. While implementing such a plan, the robot may find that it is not possible to conduct a current step (for example, the corresponding location may become inaccessible due to weather conditions). In this case, the robot needs to adopt another plan (that is, to use another word-representant) and continue its execution from the current state. The choice of alternative plans would normally be restricted by extra conditions (for example, the robot's limited power), and the order on word-representants should be helpful in picking such a plan.

Closely related to preference orders is the study of the existence of words representing a graph when extra restrictions are imposed. Such studies are also interesting in their own right from the point of view of combinatorics on words. For example, we may require that each factor of given length m contains at least one occurrence of each letter involved in the word, or that the number of letters between two consecutive equal letters is at least/most ℓ etc. A more general setup here is manifested in the following problem.

Problem 7.2.10. Given a word-representable graph G. Does there exist a word-representant for G that does not contain each word from a forbidden set of words as a factor (or as a subsequence).

A particular direction of research suggested by Problem 7.2.10 is the square-free word-representants discussed in Subsection 7.1.3.

7.3 Line Graphs

Most of the open problems in this section are from [94].

Recall from Subsection 5.4.7 that the line graph of a word-representable graph is sometimes word-representable, sometimes not, while the line graphs of non-word-representable graphs, in the cases we know, are never word-representable. Indeed, the line graphs of odd wheels, including W_5, are non-word-representable by Theorem 5.5.4, while non-word-representability of the line graphs of all graphs in Figure 3.9 was checked by Herman Z.Q. Chen [34].

Problem 7.3.1. Is the line graph of a non-word-representable graph always non-word-representable?

For a graph G define $\xi(G)$ to be the smallest non-negative integer such that $L^k(G)$ is non-word-representable for all $k \geqslant \xi(G)$. Theorem 5.5.6 shows that $\xi(G) \leq 4$ for a graph which is neither a path, nor a cycle, nor the claw $K_{1,3}$, while paths, cycles and the claw have $\xi(G) = +\infty$.

Problem 7.3.2. What is the distribution of the statistic ξ? In particular, for a given size of graphs (that is, fixing the number of vertices), can we compare the number of graphs G with $\xi(G) = i$ and $\xi(G) = j$ for $i \neq j$?

Note that for answering the second question in Problem 7.3.2 we may not need to explicitly find the cardinalities of the sets involved, possibly providing instead some kind of embedding to compare the cardinalities.

Recall that [21] provides a characterization of line graphs in terms of nine particular forbidden induced subgraphs. It would be nice to establish a similar result in the context of word-representability.

Problem 7.3.3. Is there a finite characterization of the class of word-representable line graphs in terms of minimal forbidden induced subgraphs?

7.4 Crown Graphs

Recall the definition of a crown graph $H_{k,k}$ in Definition 4.2.4 and see Figure 7.4 for a few small such graphs. It is a well known fact that the dimension of the poset corresponding to $H_{k,k}$ is k for $k \geq 2$, and thus $H_{k,k}$ is a k-comparability graph (it is permutationally k-representable but not permutationally $(k-1)$-representable). In Lemma 4.2.9, the following way to permutationally represent $H_{k,k}$ was suggested. For $k \geq 2$, concatenate the permutation $12\cdots(k-1)k'k(k-1)'\cdots 2'1'$ together with all permutations obtained from this by simultaneous exchange of k and k' with m and m', respectively, for $m = 1, \ldots, k-1$. See Table 7.1 for permutational k-representations of $H_{k,k}$ for $k = 1, 2, 3, 4$, where $k = 1$ is the special case.

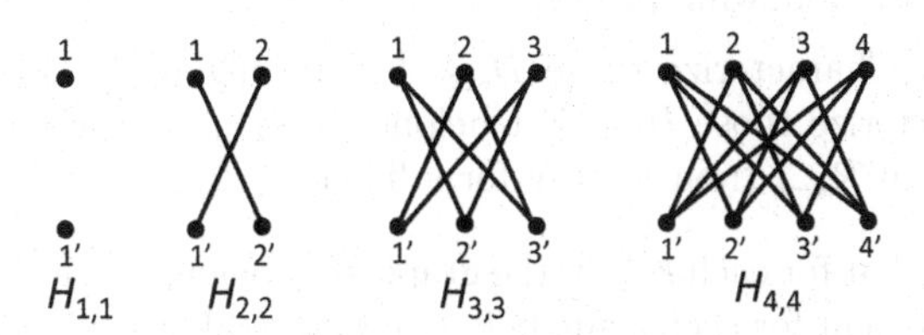

Figure 7.4: The crown graph $H_{k,k}$ for $k = 1, 2, 3, 4$

Consider the cycle graph C_5 on five edges. C_5 is in $\mathcal{R}_2$ (it is 2-word-representable, but not 1-word-representable; see Section 3.2), but it is not a comparability graph and thus, by Theorem 3.4.3, C_5 is not permutationally representable. On the other hand, the crown graphs $H_{1,1}$ and $H_{2,2}$, being 2-comparability graphs, belong to $\mathcal{R}_2$. Also, $H_{3,3}$, being a 3-comparability graph, belongs to $\mathcal{R}_2$, since $H_{3,3}$ is the cycle graph C_6. Moreover, $H_{4,4}$, being a 4-comparability graph, belongs to $\mathcal{R}_3$, which follows from the fact that $H_{4,4}$ is the prism Pr_4 (the three-dimensional cube) and Theorem 5.2.19 can be applied.

k	permutational k-representation of the crown graph $H_{k,k}$
1	$11'1'1$
2	$12'21'21'12'$
3	$123'32'1'132'23'1'231'13'2'$
4	$1234'43'2'1'1243'34'2'1'1342'24'3'1'2341'14'3'2'$

Table 7.1: Permutational k-representation of the crown graph $H_{k,k}$ for $k = 1, 2, 3, 4$

Crown graphs, being bipartite graphs and thus comparability graphs, provide an interesting case study of relations between k-comparability graphs and k-word-representable graphs. While each k-comparability graph is necessarily k-word-representable, in some cases such a graph is also $(k-1)$-word-representable, and, in fact, it is in $\mathcal{R}_{k-1}$ in the situations known to us. Thus, it seems that by giving up permutational representability we should be able to come up with a shorter representation of a given comparability graph. However, we do not know whether this is essentially always the case (except for some particular cases like the graphs $H_{1,1}$ and $H_{2,2}$). Thus we state the following open problem.

Problem 7.4.1. Characterize k-comparability graphs that belong to $\mathcal{R}_{k-\ell}$ for a fixed ℓ. In particular, characterize those k-comparability graphs that belong to $\mathcal{R}_{k-1}$. Is the set of k-comparability graphs that belong to $\mathcal{R}_{k-\ell}$ (non-)empty for a fixed $\ell \geq 2$?

A step towards solving Problem 7.4.1 might be to first understand crown graphs and solve the following problem.

Problem 7.4.2. Characterize those $H_{k,k}$ that belong to $\mathcal{R}_{k-\ell}$ for a fixed ℓ. In particular, characterize those $H_{k,k}$ that belong to $\mathcal{R}_{k-1}$. Is the set of crown graphs $H_{k,k}$ that belong to $\mathcal{R}_{k-\ell}$ (non-)empty for a fixed $\ell \geq 2$?

We suspect that for each $k \geq 3$, the graph $H_{k,k}$ belongs to $\mathcal{R}_{k-1}$. If we managed to prove this statement for even a single $k \geq 5$, we would answer at once (negatively) the question in Problem 7.2.7. A step towards solving Problem 7.4.2 might be the following observation: take the concatenation of the permutations representing $H_{k,k}$ discussed above and remove the leftmost $k-1$ letters $1, 2, \ldots, k-1$ and the rightmost $k-1$ letters $2', 3', \ldots, k'$. The remaining word will still represent $H_{k,k}$, which is not difficult to see. One would then need to remove from the word two more letters, one copy of k and one copy of $1'$, then possibly rearrange of the remaining letters in the hope of obtaining a word that $(k-1)$-word-represents $H_{k,k}$. For example, for $k = 3$, we begin with the word representing $H_{3,3}$ in Table 7.1, and remove its first two and last two letters to obtain the following representation of $H_{3,3}$:

$$3'32'1'132'23'1'231'1.$$

Now, remove the leftmost 3 and the leftmost $1'$, make the last letter 1 into the first letter 1 (that is, perform the cyclic one-position shift to the right), and, finally, replace the six rightmost letters in the obtained word by the word obtained by listing these letters in the reverse order to obtain a 2-word-representation of $H_{3,3}$ (this is the same representation as one would obtain for the cycle C_6 by following the strategy described in Section 3.2): $13'2'132'1'321'3'2$.

Unfortunately, similar steps do not work for 3-word-representing $H_{4,4}$, and it is not so clear which copies of the unwanted letters 4 and $1'$ one should remove before conducting further rearrangements. In either case, even if one could demonstrate how to obtain a $(k-1)$-word-representation of $H_{k,k}$ for some $k \geq 5$, it is not so clear what generic argument would show that no $(k-2)$-word-representation exists, if this were the case.

7.5 Triangulations of Polyominoes

The following directions of research, to extend the studies discussed in Sections 5.6 and 5.7, were offered in [68].

- Does Theorem 5.7.1 hold if we allow more than one domino tile? Note that using the current approach, the analysis involved seems to require too many cases to be considered. In either case, the problem has the following particular subproblem:
  - Does Theorem 5.7.1 hold if we allow just horizontal domino tiles (equivalently, just vertical domino tiles)?
- Does Theorem 5.7.1 hold if the domino tile is placed on other, not necessarily rectangular, convex polyominoes? What about allowing more than one domino tile to be used? Note that the counterexample in Figure 5.31 can be used to show that Theorem 5.7.1 does not hold for non-convex polyominoes — see Figure 7.5.

7.6 Approaches to Tackle Problems on Representing Graphs by Words

There are several basic approaches, some standard, some novel, to deal with word-representable graphs.

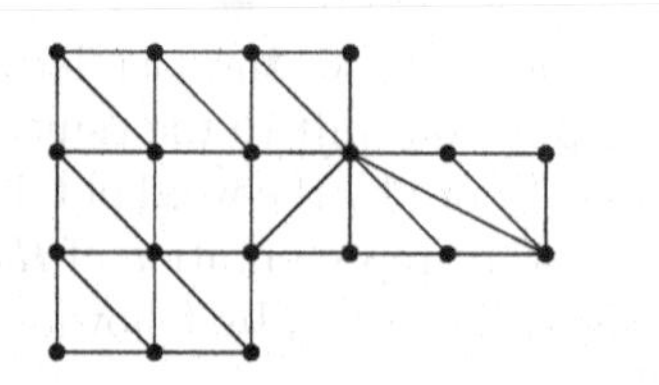

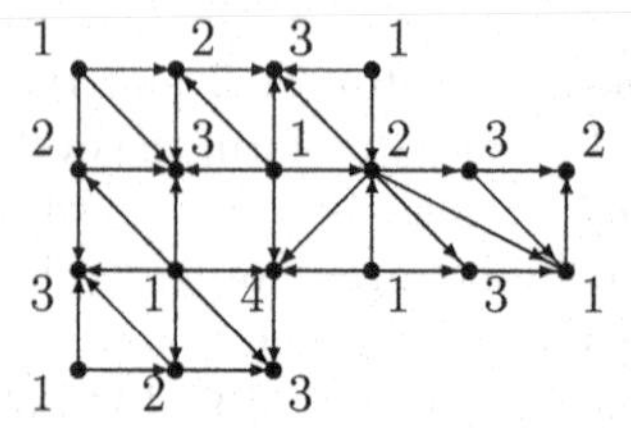

Figure 7.5: A non-3-colourable triangulation of a polyomino with a single domino tile, and its semi-transitive orientation

- A *graph-theoretic approach* is based on the characterization theorem on word-representation in terms of semi-transitive orientations. Namely, given a graph, to see whether it is word-representable or not, one could try to construct a semi-transitive orientation, or to prove that such an orientation does not exist (as is done, e.g. in the proof of Theorem 5.6.2). This approach is suitable for tackling Problem 7.1.11.

  One could also try to colour a given graph using at most three colours, succeeding in which would show that the graph is word-representable. This approach can be applied to the Petersen graph. We expect this approach to be useful in Problem 7.1.11. On the other hand, to justify non-word-representability of a given graph, one may start looking for a non-comparability neighbourhood of a vertex, which would instantly give non-word-representability of the graph, as is done in proving that the wheel W_5 is non-word-representable (see Subsection 3.5.1). This approach could be useful in Problems 7.1.1 and 7.1.11.

  Another relevant approach is to utilize known results in graph theory. For example, the fact that almost all graphs contain W_5 as an induced subgraph shows that almost all graphs are non-word-representable. There is a chance to use this approach in Problems 7.1.1 and 7.1.3.

- A *combinatorics on words approach* involves manipulations of sequences, either building an explicit word-representation of a given graph, as is done, for example, in 3-word-representing an arbitrary prism, or proving that such a representation does not exist, as is done in showing, for example, that the Petersen graph is not 2-word-representable (see Theorem 5.2.1). A useful observation here, given in Proposition 3.2.7, is that a cyclic shift of a word k-word-representing a graph also k-word-represents the graph. This approach should be basic for solving Problems 7.1.2, 7.1.4, 7.1.10, 7.2.9 and 7.2.10.

- One can use known results on word-representable graphs as building blocks to obtain new results. It is helpful here to know operations on graphs respecting (non)-word-representability. For example, given two word-representable graphs, one can connect them by an edge whose end points are arbitrary vertices in the graphs and still get a word-representable graph (see Theorem 5.4.1). The same holds for gluing two word-representable graphs at a vertex (see Theorem 5.4.2). The last observation should be helpful in the situation when a graph under consideration has a cut-vertex (that is, a vertex whose deletion, together with the edges incident to it, leads to a graph with two or more connected components).

- A *computer-based approach* relies on an exhaustive search, as was done, for example, in finding 3-representations of the Petersen graph (see Theorem 5.2.1). Of great importance is the result allowing the consideration of only uniform word-representations. A typical task here is to eliminate symmetries in order to be able to go through all possibilities in reasonable time to establish a desired property. We can think of using computer-based approaches (by arranging appropriate experiments) in all of the problems, possibly except for Problem 7.1.1.

- A *reduction to a known result.* Approaching some of the problems, one could try to use known facts, as was already mentioned above. Examples of when this approach works are Theorems 4.2.16 and 4.2.10; the latter was obtained by reducing the problem to a known result on partially ordered sets. Similarly, Theorem 3.4.3 is a reformulation in our language of a known result on total orders.

7.7 Other Notions of Word-Representable Graphs

Variations of the notion of word-representability of graphs have been studied in the literature. For example, given a word, we may impose the existence of an edge in the graph if and only if at least two occurrences of the corresponding letters in the word alternate (in our setup, all occurrences must alternate); the corresponding class of graphs, *polygon-circle graphs*, has a simple geometric definition in terms of polygons inscribed in a circle, and we discuss it in Section 8.6 below.

Chapter 6 provided a far-reaching generalization of the notion of a word-representable graph. As is mentioned in that chapter, apart from our main generalization, given in Definition 6.1.9, of the notion of a word-representable graph, we have another generalization, given in Definition 7.7.19 below. We also state below some other ways to define the notion of a (directed) graph representable by words. Our definitions can be generalized to the case of *hypergraphs* by simply allowing words defining edges/non-edges over alphabets containing more than just two letters; this would be very similar in nature to alternation word digraphs introduced in Subsection 3.6.1. However, in this book we do not consider hypergraphs.

We note that none of the notions introduced in this section (all of them came from [84]) have been studied to date. Thus, the current section provides plenty of material for further research.

7.7.1 Occurrences, Exact Occurrences and Exact Matches

Definition 7.7.1. Given a word $u = u_1 \cdots u_j \in \mathbb{P}^*$ such that $\mathrm{red}(u) = u$, and a word $w = w_1 \cdots w_n \in \mathbb{P}^*$, we say that u *occurs* in w as a pattern if there exist $1 \leq i_1 < \cdots < i_j \leq n$ such that $\mathrm{red}(w_{i_1} \cdots w_{i_j}) = u$, and that w *avoids* u if u does not occur in w.

Example 7.7.2. The pattern 121 occurs in the word 376134 twice (the subsequences 373 and 363 are its occurrences).

Remark 7.7.3. The notion of an occurrence of a pattern in a word in Definition 7.7.1 is known as an occurrence of a *classical pattern* in the theory of patterns in permutations and words [88]. This notion for permutations is introduced in Definition 8.3.1 below.

Note that a word w contains an occurrence of a pattern u if and only if its reverse $r(w)$ (defined in Definition 3.0.12) contains an occurrence of the pattern $r(u)$.

Definition 7.7.4. Given a word $v = v_1 \cdots v_j \in \mathbb{P}^*$ and a word $w = w_1 \cdots w_n \in \mathbb{P}^*$, we say that v *exactly occurs* in w if there exist $1 \leq i_1 < \cdots < i_j \leq n$ such that $w_{i_1} \cdots w_{i_j} = v$ and that w *exactly avoids* v if v does not exactly occur in w. We also say that w has an *exact v-match starting at position i* if $w_i w_{i+1} \cdots w_{i+j-1} = v$.

Example 7.7.5. Suppose that $w = 286843341$. Then the word 633 exactly occurs in w, while w exactly avoids the word 2244. Also, w has an exact 684-match but no exact 351-match.

7.7.2 Dealing with Sets of Words

In this section, some notions that have previously been defined for words are extended to sets of words. More precisely, Definitions 7.7.6 and 7.7.8 below are generalizations of Definitions 6.1.5 and 7.7.4, respectively.

Definition 7.7.6. Let Γ be a set of words in $\mathbb{P}^*$ such that $\mathrm{red}(u) = u$ for all $u \in \Gamma$. Then we say that Γ *occurs* in $w = w_1 \cdots w_n \in \mathbb{P}^*$ if there exist $1 \leq i_1 < \cdots < i_j \leq n$ such that $\mathrm{red}(w_{i_1} \cdots w_{i_j}) \in \Gamma$, and that w *avoids* Γ if Γ does not occur in w. We also say that w has a *Γ-match starting at position i* if $\mathrm{red}(w_i w_{i+1} \cdots w_{i+j-1}) \in \Gamma$.

Example 7.7.7. The set $\Gamma = \{3121, 112\}$ occurs in the word $w = 243124$ (because $\text{red}(224) = 112$), while the word 12431 avoids Γ. Also, the word 4336418 has a Γ-match (because $\text{red}(336) = 112$) starting at position 2.

Definition 7.7.8. If Δ is any set of words in $\mathbb{P}^*$, we say that Δ *exactly occurs* in $w = w_1 \cdots w_n \in \mathbb{P}^*$ if there exist $1 \leq i_1 < \cdots < i_j \leq n$ such that $w_{i_1} \cdots w_{i_j} \in \Delta$, and that w *exactly avoids* Δ if Δ does not occur in w. We also say that w has an *exact Δ-match starting at position i* if $w_i w_{i+1} \cdots w_{i+j-1} \in \Delta$.

Example 7.7.9. The set $\Delta = \{312, 414\}$ exactly occurs in the word $w = 243142$ (because of the word 312), while 431556 exactly avoids Δ. Also, the word 11241423 has an exact Δ-match (because of the word 414) starting at position 4.

7.7.3 Representing Graphs

We defined the notion of a u-representable graph in Definition 6.1.9. More generally, we can make the same definition for sets of words.

Definition 7.7.10. Let Γ be a set of words in $\{1,2\}^*$ such that $\text{red}(u) = u$ for all $u \in \Gamma$. Then we say that a graph $G = (V, E)$, where $V \subset \mathbb{P}$, is *Γ-representable* if there exists a word $w \in \mathbb{P}^*$ such that $A(w) = V$ and for all $x, y \in V$, $xy \notin E$ if and only if $w_{\{x,y\}}$ has a Γ-match.

Example 7.7.11. The word 4332414 $\{12, 211\}$-represents the left graph in Figure 7.6.

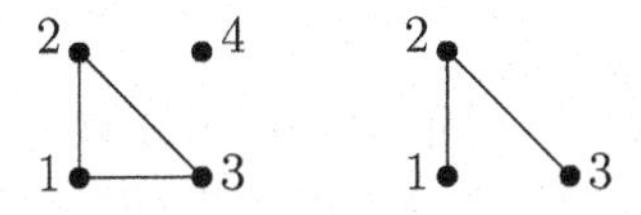

Figure 7.6: Two graphs

Problem 7.7.12. Study Γ-representable graphs for various (natural) sets Γ. In particular, describe the class of $\{12, 211\}$-representable graphs.

Definition 7.7.13. Let Γ be a set of words in $\{1,2\}^*$ such that $\text{red}(u) = u$ for all $u \in \Gamma$. Then we say that a graph $G = (V, E)$, where $V \subset \mathbb{P}$, is *Γ-occurrence representable* if there exists a word $w \in \mathbb{P}^*$ such that $A(w) = V$ and for all $x, y \in V$, $xy \notin E$ if and only if Γ occurs in $w_{\{x,y\}}$.

Example 7.7.14. The word 311322 $\{121, 212\}$-occurrence represents the right graph in Figure 7.6.

Problem 7.7.15. Study Γ-occurrence representable graphs for various natural/small sets Γ. In particular, describe the class of $\{121, 212\}$-representable graphs.

Definition 7.7.16. In the case where $\Gamma = \{u\}$ consists of a single word, we simply say that a graph G is u-occurrence representable if G is Γ-occurrence representable.

Example 7.7.17. The 11-occurrence representable graphs are very simple. That is, if a word $w = w_1 \cdots w_n$ 11-occurrence represents a graph $G = (V, E)$, then any vertex x such that w has two or more occurrences of x cannot be connected to any other vertex y since 11 will always occur in $w_{\{x,y\}}$. Let $I = \{x \in V : x \text{ occurs more than once in } w\}$ and $J = \{y \in V : y \text{ occurs exactly once in } w\}$. Then it is easy to see that the elements of J must form a clique in G, while the elements of I form an edgeless graph. Thus, if G is 11-occurrence representable, then G consists of a clique together with a set of isolated vertices. Clearly, all such graphs are 11-occurrence representable, which gives a characterization of 11-occurrence representable graphs.

A straightforward observation is recorded in the following proposition.

Proposition 7.7.18. *([84]) The sets of* 12*-representable graphs and* 12*-occurrence representable graphs coincide.*

Proof. It is not difficult to see that a word contains a 12-match if and only if it contains a 12-occurrence. □

Similarly, we have the following analogues of our definition for exact matchings and exact occurrences.

Definition 7.7.19. Let Δ be a set of words in $\mathbb{P}^*$. Then we say that a graph $G = (V, E)$, where $V \subset \mathbb{P}$, is *exact-Δ-representable* if there is a word $w \in \mathbb{P}^*$ such that $A(w) = V$ and for all $x, y \in V$, $xy \notin E$ if and only if $w_{\{x,y\}}$ has an exact Δ-match.

Example 7.7.20. The word 41424234 exact-$\{14, 24, 34\}$-represents the left graph in Figure 7.6.

Problem 7.7.21. Study exact-Δ-representable graphs for various (natural/small) sets Δ.

Definition 7.7.22. Let Δ be a set of words in $\mathbb{P}^*$. Then we say that a graph $G = (V, E)$, where $V \subset \mathbb{P}$, is *exact-Δ-occurrence representable* if there is a word $w \in \mathbb{P}^*$ such that $A(w) = V$ and for all $x, y \in V$, $xy \notin E$ if and only if Δ exactly occurs in $w_{\{x,y\}}$.

Example 7.7.23. The word 13211 exact-$\{131, 212\}$-occurrence represents the right graph in Figure 7.6.

Note that to avoid trivialities, while dealing with exact matchings or occurrences, the sets of words defining (non-)edges should be large and hopefully contain at least one word for each pair of vertices in V.

Problem 7.7.24. Study exact-Δ-occurrence representable graphs for various (natural/small) sets Δ.

7.7.4 Simple Results on the New Definitions

It is easy to see that we have an analogue of Proposition 3.0.8 for each of the notions introduced in Subsection 7.7.3. That is, we have the following statements.

Proposition 7.7.25. *Let Γ be a set of words in $\mathbb{P}^*$ such that* $\mathrm{red}(u) = u$ *for all* $u \in \Gamma$ *and let H be an induced subgraph of G.*

1. *If G is Γ-representable, then H is Γ-representable.*
2. *If G is Γ-occurrence representable, then H is Γ-occurrence representable.*

Proposition 7.7.26. *Let Δ be a set of words in $\mathbb{P}^*$ and H be a subgraph of G.*

1. *If G is exact-Δ-representable, then H is exact-Δ-representable.*
2. *If G is exact-Δ-occurrence representable, then H is exact-Δ-occurrence representable.*

Recall the definitions of the reverse $r(u)$ in Definition 3.0.12 and the complement $c(u)$ in Definition 6.1.10.

Definition 7.7.27. If Γ is a set of words in $\mathbb{P}^*$, then we let $r(\Gamma) = \{r(u) : u \in \Gamma\}$.

Example 7.7.28. If $\Gamma = \{13, 2441, 552, 121\}$ then $r(\Gamma) = \{31, 1442, 255, 121\}$.

Definition 7.7.29. If Δ is a set of words $u \in \mathbb{P}^*$ such that $\mathrm{red}(u) = u$, then we let $c(\Delta) = \{c(u) : u \in \Delta\}$.

Example 7.7.30. If $\Delta = \{1332, 43142, 12\}$ then $c(\Delta) = \{3112, 12413, 21\}$.

We have the following statement generalizing and extending Proposition 6.1.12.

Proposition 7.7.31. *([84]) Let G be a graph, Γ be a set of words in $\mathbb{P}^*$ such that* $\text{red}(u) = u$ *for all* $u \in \Gamma$. *Then*

1. *G is Γ-representable if and only if G is $r(\Gamma)$-representable.*
2. *G is Γ-occurrence representable if and only if G is $r(\Gamma)$-occurrence representable.*

Proof. It is easy to see that if w witnesses that G is Γ-representable (resp., Γ-occurrence representable), then $r(w)$ witnesses that G is $r(\Gamma)$-representable (resp., $r(\Gamma)$-occurrence representable). □

Recall the definition of the supplement $c(G)$ of a graph G given in Definition 6.1.13. The following statement generalizes and extends Proposition 6.1.14.

Proposition 7.7.32. *([84]) Let G be a graph, and Δ be a set of words in $\mathbb{P}^*$ such that* $\text{red}(u) = u$ *for all* $u \in \Delta$. *Then*

1. *G is Δ-representable if and only if $c(G)$ is $c(\Delta)$-representable.*
2. *G is Δ-occurrence representable if and only if $c(G)$ is $c(\Delta)$-occurrence representable.*

Proof. It is easy to see that if w witnesses that G is Δ-representable (resp., Δ-occurrence representable), then $c(w)$ witnesses that $c(G)$ is $c(\Delta)$-representable (resp., $c(\Delta)$-occurrence representable). □

7.7.5 Wilf-Equivalences

The notion of Wilf-equivalence is important in the theory of patterns in permutations and words [88]. In that theory, two sets of patterns are Wilf-equivalent if the number of permutations of length n avoiding the first set is the same as that avoiding the second set for all $n \geq 0$. Here we provide analogues of this notion in the theory of graphs representable by words.

Definition 7.7.33. Given two words $u, v \in \mathbb{P}^*$, we say that u and v are *matching-representation Wilf-equivalent* if any u-representable graph G is also v-representable, and vice versa.

Definition 7.7.34. We say that u and v are *occurrence-representation Wilf-equivalent* if any u-occurrence representable graph G is also v-occurrence representable, and vice versa.

Our propositions in Subsection 7.7.4 show that the matching-representation and occurrence-representation Wilf-equivalence classes are closed under reverse and complement.

Problem 7.7.35. Give examples of matching-representation and/or occurrence-representation Wilf-equivalent words that are not obtained from each other by applying the operation of reverse or complement, or any composition of those. Can one describe any necessary and/or sufficient conditions for two words to be Wilf-equivalent?

7.7.6 Representing Directed Graphs

The idea of using patterns to represent graphs can also be extended to give us a notion of representing directed graphs via words, which is not a focus of this book though, and thus there is very little discussion of this topic.

For the definitions in this subsection, we assume that we are given a directed graph $G = (V, E)$, where $V \subset \mathbb{P}$ and $E \subset V \times V$, and we are given two sets of words Γ, Δ in $\mathbb{P}^*$ such that $\text{red}(u) = u$ for all $u \in \Gamma$ and $\text{red}(v) = v$ for all $v \in \Delta$. We also assume that G has no loops or multiple edges with the same orientation.

Definition 7.7.36. We say that a directed graph $G = (V, E)$ is (Γ, Δ)-*matching representable* if there is a word $w \in \mathbb{P}^*$ such that $A(w) = V$ and for all pairs $x < y$ in V, $x \to y \notin E$ if and only if $w_{\{x,y\}}$ has a Γ-match and $y \to x \notin E$ if and only if $w_{\{x,y\}}$ has a Δ-match.

Example 7.7.37. The word 43324414432 $(\{122, 211\}, \{212\})$-matching-represents the left graph in Figure 7.7.

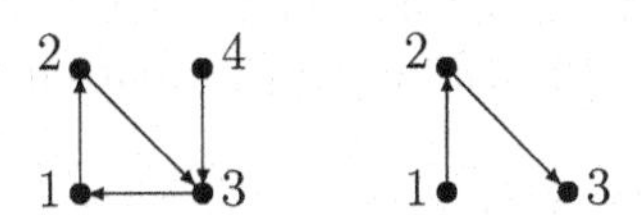

Figure 7.7: Two directed graphs

Problem 7.7.38. Study (Γ, Δ)-representable graphs for various (natural/small) sets Γ and Δ.

Definition 7.7.39. We say that a directed graph $G = (V, E)$ is (Γ, Δ)-*occurrence representable* if there is a word $w \in \mathbb{P}^*$ such that $A(w) = V$ and for all pairs $x < y$ in V, $x \to y \notin E$ if and only if Γ occurs in $w_{\{x,y\}}$ and $y \to x \notin E$ if and only if Δ occurs in $w_{\{x,y\}}$.

Example 7.7.40. The word 311322 $(\{121, 212\}, \{11\})$-occurrence represents the right graph in Figure 7.7.

Problem 7.7.41. Study (Γ, Δ)-occurrence representable graphs for various (natural/small) sets Γ and Δ.

We can give similar definitions for exact matching and exact occurrences.

Definition 7.7.42. We say that a directed graph $G = (V, E)$ is *exact-*(Γ, Δ)-*matching representable* if there is a word $w \in \mathbb{P}^*$ such that $A(w) = V$ and for all pairs $x < y$ in V, $x \to y \notin E$ if and only if $w_{\{x,y\}}$ has an exact Γ-match and $y \to x \notin E$ if and only if $w_{\{x,y\}}$ has an exact Δ-match.

Example 7.7.43. The word 2213324 exact-$(\{4\}, \{21, 32, 223, 14, 24\})$-matching- represents the left graph in Figure 7.7.

Problem 7.7.44. Study exact-(Γ, Δ)-representable graphs for various (natural) sets Γ and Δ.

Definition 7.7.45. We say that a directed graph $G = (V, E)$ is *exact-*(Γ, Δ)-*occurrence representable* if there is a word $w \in \mathbb{P}^*$ such that $A(w) = V$ and for all pairs $x < y$ in V, $x \to y \notin E$ if and only if Γ exactly occurs in $w_{\{x,y\}}$ and $y \to x \notin E$ if and only if Δ exactly occurs in $w_{\{x,y\}}$.

Example 7.7.46. The word 123123 exact-$(\{133\}, \{1, 2, 3\})$-occurrence represents the right graph in Figure 7.7.

Problem 7.7.47. Study exact-(Γ, Δ)-occurrence representable graphs for various (natural/small) sets Γ and Δ.

Remark 7.7.48. We can obtain other notions of word-representability by mixing Γ-matches, exact Γ-matches, Γ-occurrences and exact Γ-occurrences with Δ-matches, exact Δ-matches, Δ-occurrences and exact Δ-occurrences in the definitions above.

7.8 Graphs Described by a Given Language

This section offers a general program of study of word-representable graphs. This program was offered by the first author of this book during his plenary talk at the international Permutation Patterns Conference at the East Tennessee State University, Johnson City in 2014.

A *language* is a set of words over some alphabet. One can consider the following direction of research.

A general program of research

Input: Given a language defined, e.g. through forbidden patterns studied in the theory of patterns in permutations and words [88].

Goal: Study the class of graphs defined by the language via alternations of letters in words, or, more generally via pattern-avoiding words defined in Chapter 6 or in Section 7.7. In particular, characterize the graphs that can be represented by the language.

As a trivial example related to our usual word-representability, we can consider as the input the set of all words avoiding the pattern 21 (see Definition 7.7.1), that is, words of the form

$$11\cdots 122\cdots 233\cdots 3\cdots nn\cdots n,$$

where each letter in $\{1,2,\ldots,n\}$ is assumed to appear at least once. Clearly, if we have at least two occurrences of a letter, then the vertex corresponding to this letter is isolated. On the other hand, if we have exactly one occurrence of a letter, then the corresponding vertex is adjacent to each non-isolated vertex. Thus, for a given n, there are $n+1$ (unlabelled) graphs represented by this language, each having a clique of size k, $0 \leq k \leq n$, and an independent set of size $n-k$.

As a specialization of the general program stated above we have the following direction related to pattern-avoiding permutations (avoidance in the sense of Definition 7.7.1, or in any other sense).

A program of research

Input: Given a set S of pattern-avoiding permutations.

Goal: Study the class of graphs defined by concatenation of permutations from S via alternations of letters in words, or, more generally, via pattern-avoiding words defined in Chapter 6 or in Section 7.7.

Avoidance of a two-letter pattern is trivial in this context. Indeed, only one permutation of each size (either the increasing or decreasing one depending on the pattern) avoids the pattern. Thus, concatenations of such permutations can only give word-representation of cliques.

On the other hand, even in the case of three-letter patterns the situation is far from being trivial. For example, suppose that S is the set of all permutations avoiding the pattern 132 (such permutations are counted by the *Catalan numbers*

[88] defined in Appendix B, and are well-studied in the literature). It is not difficult to see that all graphs on at most three vertices can be word-represented using concatenation of 132-avoiding permutations. For (unlabelled) graphs on four vertices, such representations are given in Figure 7.8. Note that labelling vertices in the graph can be essential in some cases, as one labelling leads to the existence of a proper representation, while another labelling may not have a proper representation (this situation is similar to u-representable graphs defined in Definition 6.1.9).

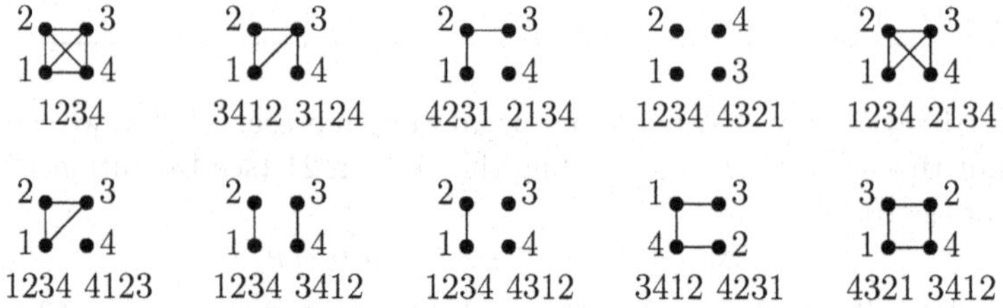

1234   3412 3124   4231 2134   1234 4321   1234 2134

1234 4123   1234 3412   1234 4312   3412 4231   4321 3412

Figure 7.8: Representing graphs by permutations avoiding the pattern 132. Note that the permutations are separated by a space for more convenient visual representation

When dealing with concatenation of pattern-avoiding permutations, by Theorem 3.4.3, we can only define a subset of comparability graphs.

Problem 7.8.1. Describe the subset of comparability graphs defined by concatenation of 132-avoiding permutations. What about 123-avoiding permutations? More generally, describe such subsets for p-avoiding permutations where p is another (longer) pattern.

Chapter 8

Interrelations Between Words and Graphs in the Literature

In previous chapters we have seen several connections between graphs and words distinct from word-representable graphs, the main topic in this book. For instance, we use words to describe graphs when we need to represent them in computer memory. On the other hand, graphs can be used to represent permutations in order to reveal some of their structural properties. The literature contains many more examples of interrelations between words and graphs and we explore some of them in this chapter.

8.1 Graphs and Sequences

8.1.1 Prüfer Sequences for Trees

By definition, a *tree* is a connected graph without cycles. Trees play an important role in graph theory and enjoy many nice properties. In particular, they are minimal connected graphs (in the sense that deletion of any edge from a tree disconnects it) and maximal graphs without cycles (in the sense that addition of any edge to a tree leads to a cycle).

In the present section, we describe an important representation of labelled trees by means of sequences of numbers proposed by Prüfer [120] in 1918. To define the notion of a Prüfer sequence, we need the following two properties of trees, where by *leaves* we mean vertices of degree 1.

- *Every tree T with at least two vertices has at least two leaves.* To see this, consider a longest chordless path P_k in T. The endpoints of P_k cannot have more than one neighbour outside of the path (else it is not longest) and cannot have neighbours on the path (else T has a cycle). Therefore, both endpoints of P_k are leaves in T.

- *A tree with n vertices has $n-1$ edges.* Indeed, for $n=1$ the statement is trivial. If T is a tree with $n \geq 2$ vertices, then by deleting any leaf from T, we obtain a tree, which is easy to see, and the result follows by induction.

Let T be a labelled tree with n vertices, i.e. a tree whose vertices are numbered by $1, 2, \ldots, n$. Let a_1 be the least leaf in T (i.e. the leaf with the least label) and let b_1 be the only neighbour of a_1. By deleting a_1 from T we obtain a new tree T_1. In this tree we find the least leaf a_2 and its neighbour b_2. By deleting a_2 from T_1 we obtain a new tree T_2. Proceeding in this way, in $n-2$ steps we obtain a tree T_{n-2} consisting of a single edge $a_{n-1}b_{n-1}$. We observe that the sequence a_1b_1, $a_2b_2, \ldots, a_{n-1}b_{n-1}$ contains all the edges of T.

Definition 8.1.1. The sequence $P(T) = b_1b_2 \cdots b_{n-2}$ arising from the above procedure is called the *Prüfer sequence* (or *Prüfer code*) of T.

Example 8.1.2. Figure 8.1 gives an example of a tree whose Prüfer sequence is 43114.

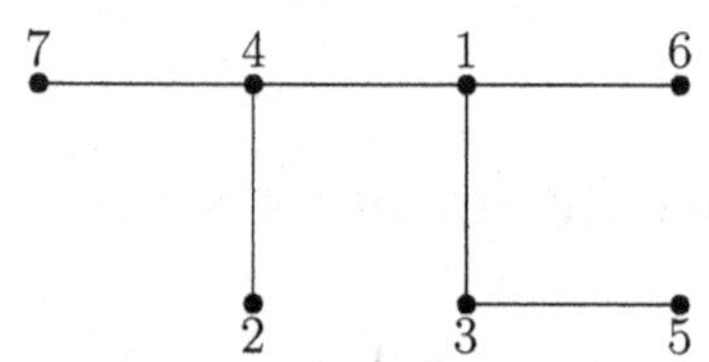

Figure 8.1: The tree with Prüfer sequence 43114

Prüfer sequences are important because they encode trees, as is shown by the following well known theorem.

Theorem 8.1.3. *The mapping P assigning to every labelled tree T its Prüfer sequence $P(T)$ is a bijection.*

Proof. First, we claim that P is injective, i.e. there is a unique labelled tree T with the sequence $P(T)$. To prove this, we need to show that, given $P(T) = b_1b_2 \cdots b_{n-2}$, one can determine the numbers $a_1, a_2, \ldots, a_{n-1}, b_{n-1}$ that uniquely describe the tree T. Let us start with a_1. We claim that

(1) *vertices with labels less than* a_1 *are not leaves in* T, which follows from the choice of a_1.

(2) *every vertex which is not a leaf appears in* $P(T)$. Indeed, if x is not a leaf, then it is adjacent to at least two other vertices. The procedure of constructing $P(T)$ finishes when only two vertices of T are left. Therefore, in the course of this construction, at least one vertex adjacent to x will be deleted. If this happens in step i, then $b_i = x$.

From (1) and (2) it follows that a_1 is the least natural number which does not appear in $P(T)$. Now assume that we have found the numbers $a_1, a_2, \ldots, a_{i-1}$. By deleting these vertices from T we obtain the tree T_{i-1}. This tree corresponds to the subsequence $b_i, b_{i+1}, \ldots, b_{n-2}$ of $P(T)$. In the tree T_{i-1}, vertex a_i is the least leaf. Arguing as before, we conclude that a_i is the least natural number in the set $\{1, 2, \ldots, n\} - \{a_1, a_2, \ldots, a_{i-1}\}$, which does not appear among $a_1, a_2, \ldots, a_{i-1}, b_i, b_{i+1}, \ldots, b_{n-2}$. This rule allows us to find the numbers $a_1, a_2, \ldots, a_{n-2}$, which are clearly pairwise distinct. The two remaining numbers in the set $\{1, 2, \ldots, n\}$ are a_{n-1} and b_{n-1}. This completes the proof of the fact that T can be uniquely decoded from $P(T)$, i.e. P is injective.

Now we turn to proving that P is surjective. The decoding procedure described above can be applied to any sequence of numbers $b_1 b_2 \cdots b_{n-2}$ chosen from a set $\{1, 2, \ldots, n\}$ and it produces numbers $a_1, a_2, \ldots, a_{n-1}, b_{n-1}$ from the same set. From these numbers we can construct the graph G with edges a_1b_1, $a_2b_2, \ldots, a_{n-1}b_{n-1}$, and we need to show that this graph is a tree. Let us denote by G_i the graph with vertex set $\{1, 2, \ldots, n\}$ and edge set $a_ib_i, \ldots, a_{n-1}b_{n-1}$. The graph G_{n-1} has a unique edge and hence has no cycles. Suppose that the graph G_{i+1} has no cycles. According to the decoding procedure, the number a_i is different from $a_{i+1}, \ldots, a_{n-1}$ and from $b_{i+1}, \ldots, b_{n-1}$. Therefore, a_i has degree 1 in G_i and hence the cycle freeness of G_{i+1} implies cycle freeness of G_i. By induction, we conclude that $G = G_1$ contains no cycles. It remains to show that G is connected.

Let k be the number of connected components in G, and n_i the number of vertices in the ith component. Since each connected component is a tree, the ith component has $n_i - 1$ edges. Therefore, the total number of edges in G is $\sum_{i=1}^{k}(n_i - 1) = n - k$. On the other hand, by definition, the number of edges in G is $n - 1$. Therefore, $k = 1$, i.e. G is connected. □

From the above theorem and the obvious fact that the number of sequences of length $n - 2$ constructed from the elements of an n-element set is n^{n-2}, we derive the following well known fact.

Corollary 8.1.4. *The number of labelled trees with n vertices is n^{n-2}.*

Thus, Theorem 8.1.3 provides an alternative proof of the celebrated Cayley's formula for the number of labelled trees.

8.1.2 Threshold Graphs

Definition 8.1.5. Let G be a graph with the vertex set $\{v_1, v_2, \ldots, v_n\}$. Then $(x_1, x_2, \ldots, x_n)$ is the *characteristic vector* of an independent set I in G if $x_i = 1$ whenever $v_i \in I$ and $x_i = 0$ otherwise.

Threshold graphs were introduced by Chvátal and Hammer in [39] as follows.

Definition 8.1.6. A graph G with n vertices is threshold if there exist real numbers $a_1, a_2, \ldots, a_n$ and b such that the zero-one solutions of

$$\sum_{j=1}^{n} a_j x_j \leq b$$

are precisely the characteristic vectors of the independent sets in G.

In the same paper, the authors proved the following characterization of threshold graphs.

Theorem 8.1.7. *A graph $G = (V, E)$ is threshold if and only if at least one of the following holds:*

(1) *G has no induced subgraphs isomorphic to $2K_2$, P_4 or C_4.*

(2) *There is an ordering $v_1, v_2, \ldots, v_n$ of G's vertices such that for each $j = 1, \ldots, n-1$, vertex v_j is adjacent either to every vertex in the set $\{v_{j+1}, \ldots, v_n\}$ or to no vertex in this set.*

(3) *There is a partition of V into disjoint sets A of size k and B, where A is an independent set and B is a clique, and an ordering $u_1, u_2, \ldots, u_k$ of vertices in A such that $N(u_1) \subseteq N(u_2) \subseteq \cdots \subseteq N(u_k)$, where $N(u)$ denotes the set of vertices adjacent to u.*

Remark 8.1.8. Observe that the linear order of A in the third characterization of threshold graphs given in Theorem 8.1.7 implies the existence of a linear order $w_1, w_2, \ldots, w_{n-k}$ of vertices in B such that $N[w_1] \subseteq N[w_2] \subseteq \cdots \subseteq N[w_{n-k}]$.

From the second characterization of Theorem 8.1.7 it follows that every n-vertex threshold graph G can be described by a binary sequence $\alpha_1, \ldots, \alpha_{n-1}$ of length $n-1$, where $\alpha_j = 1$ if v_j is adjacent to all the vertices in $\{v_{j+1}, \ldots, v_n\}$ and $\alpha_j = 0$ if v_j is adjacent to none of the vertices in $\{v_{j+1}, \ldots, v_n\}$. Moreover, it is not difficult to see that on the set of unlabelled graphs this mapping is injective, i.e. it maps non-isomorphic graphs to different sequences. Also, it is obvious that this mapping is surjective, i.e. every binary sequence represents a threshold graph.

Corollary 8.1.9. *There is a bijection between the set of unlabelled n-vertex threshold graphs and the set of binary sequences of length $n-1$.*

In spite of the simple structure of threshold graphs, this notion gave rise to a vast literature on the topic, including the book [108]. In particular, in [76] Hammer and Kelmans describe an *n-universal threshold graph*, i.e. a graph containing every threshold graph with n vertices as an induced subgraph. Not surprisingly an n-universal threshold graph contains $2n-1$ vertices, because a binary sequence of length $2n-2$ where $n-1$ zeros and $n-1$ ones strictly alternate contains every binary sequence of length $n-1$ as a subsequence.

The representation of threshold graphs as binary words shows that the set of threshold graphs is *well-quasi-ordered* (see Section 8.2 for definition) by the induced subgraph relation. More generally, the set of so-called *k-letter graphs* (considered in Section 8.2) is well-quasi-ordered by induced subgraphs for any fixed k. We discuss this topic in more detail in Section 8.2.

The same conclusion about well-quasi-orderability of threshold graphs by induced subgraphs also follows from the fact that threshold graphs are permutation graphs and the respective class of permutations constitutes a geometric grid class. This topic and related results are discuss in Section 8.3.

8.2 Letter Graphs

Let Σ be a finite alphabet and $\mathcal{P} \subseteq \Sigma^2$ a set of ordered pairs of symbols from Σ. With each word $w = w_1 w_2 \cdots w_n$ with $w_i \in \Sigma$ we associate a graph $G(\mathcal{P}, w)$, called the *letter graph* of w, by defining $V(G(\mathcal{P}, w)) = \{1, 2, \ldots, n\}$ with i being adjacent to $j > i$ if and only if the ordered pair (w_i, w_j) belongs to $\mathcal{P}$.

Example 8.2.1. Let $\Sigma = \{0, 1\}$ and $\mathcal{P} = \{(1,1), (1,0)\}$. Then for any binary word $w = w_1 w_2 \cdots w_n$, the graph $G(\mathcal{P}, w)$ is threshold. Indeed, if $w_i = 1$, then i is adjacent to every vertex $j > i$ (since both pairs $(1,1)$ and $(1,0)$ belong to $\mathcal{P}$), and if $w_i = 0$ then i is not adjacent to any vertex $j > i$ (since neither $(0,0)$ nor $(0,1)$

belong to $\mathcal{P}$). We observe that this representation is similar but not identical to the correspondence between graphs and words described in Corollary 8.1.9.

By excluding from $\mathcal{P}$ the pair $(1, 1)$, we transform the letter graph of w into a bipartite graph, because the clique formed by symbols 1 in w becomes an independent set under this transformation. According to Theorem 8.1.7(3) (and the remark after the theorem), the vertices in each part of this bipartite graph can be linearly ordered under inclusion of their neighbourhoods. Bipartite graphs possessing this property are known as *chain graphs.*

It is not difficult to see that every graph G is a letter graph in a sufficiently large alphabet with an appropriate set $\mathcal{P}$. The minimum ℓ such that G is a letter graph in an alphabet of ℓ letters is the *lettericity* of G and is denoted $\ell(G)$. A graph is a k-letter graph if its lettericity is at most k.

The notion of k-letter graphs was introduced in [115] and in the same paper the author characterized k-letter graphs as follows.

Theorem 8.2.2. *([115]) A graph G is a k-letter graph if and only if*

1. *there is a partition $V_1, V_2, \ldots, V_p$ of $V(G)$ with $p \leq k$ such that each V_i is either a clique or an independent set in G, and*
2. *there is a linear ordering L of $V(G)$ such that for each pair of distinct indices $1 \leq i, j \leq p$, the intersection of $E(G)$ with $V_i \times V_j$ is one of the following four types (where L is considered as a binary relation, i.e. as a set of pairs):*
   - *(a)* $L \cap (V_i \times V_j)$;
   - *(b)* $L^{-1} \cap (V_i \times V_j)$;
   - *(c)* $V_i \times V_j$;
   - *(d)* $\emptyset$.

Let us observe that conditions (a) and (b) of this theorem correspond to the cases when the edges between two sets V_i and V_j form a chain graph.

The importance of the notion of k-letter graphs is due to the fact that for each fixed value of k, the set of all k-letter graphs is well-quasi-ordered by the induced subgraph relation. To make things more precise, let is introduce some definitions and results from the theory of partial orders.

Detour: *well-quasi-ordering.*

For a set A, we denote by A^2 the set of all ordered pairs of (not necessarily distinct) elements from A. A *binary relation* on A is a subset of A^2. If a binary relation $\mathcal{R} \subset A^2$ is

- *reflexive*, i.e. $(a, a) \in \mathcal{R}$, and
- *transitive*, i.e. $(a, b) \in \mathcal{R}$ and $(b, c) \in \mathcal{R}$ imply $(a, c) \in \mathcal{R}$,

then $\mathcal{R}$ is a *quasi-order* (also known as *pre-order*). If additionally $\mathcal{R}$ is

- *antisymmetric*, i.e. $(a, b) \in \mathcal{R}$ and $(b, a) \in \mathcal{R}$ imply $a = b$,

then $\mathcal{R}$ is a *partial order*.

We say that two elements $a, b \in A$ are comparable with respect to $\mathcal{R}$ if either $(a, b) \in \mathcal{R}$ or $(b, a) \in \mathcal{R}$. A set of pairwise comparable elements of A is called a *chain* and a set of pairwise incomparable elements of A is called an *antichain*.

A quasi-ordered set is *well-quasi-ordered* if it contains

- neither infinite strictly decreasing chains, in which case we say that the set is *well-founded*,
- nor infinite antichains.

Well-quasi-ordering (*WQO*) is a highly desirable property and frequently discovered concept in mathematics and theoretical computer science [59, 101]. One of the most remarkable recent results in this area is the proof of Wagner's conjecture stating that the set of all finite graphs is well-quasi-ordered by the minor relation [121]. However, the subgraph or induced subgraph relation is not a well-quasi-order, since the set of all chordless cycles constitutes an infinite antichain with respect to both relations. Other examples of important relations that are not well-quasi-orders are the pattern containment relation on permutations [127], the embeddability relation on tournaments [35], minor ordering of matroids [80] and the factor (contiguous subword) relation on words [47]. On the other hand, each of these relations may become a well-quasi-order under some additional restrictions.

A simple but very powerful tool for proving well-quasi-orderability is the celebrated Higman's Lemma, which can be stated as follows. Let M be a set with a quasi-order $\leq$. We can extend $\leq$ from M to M^* as follows: $(a_1, \ldots, a_m) \leq (b_1, \ldots, b_n)$ if and only if there is an injection $f : \{a_1, \ldots, a_m\} \to \{b_1, \ldots, b_n\}$ with $a_i \leq f(a_i)$ for each $i = 1, \ldots, m$. Then Higman's Lemma from [79] states the following.

Lemma 8.2.3. *([79]) If $(M, \leq)$ is a WQO, then $(M^*, \leq)$ is a WQO.*

With the help of Higman's Lemma it is not difficult to show that for each fixed value of k, the set of all k-letter graphs is well-quasi-ordered by the induced subgraph relation. This was formally proved in [115].

Theorem 8.2.4. *([115]) For each fixed value of k, the set of all k-letter graphs is well-quasi-ordered by the induced subgraph relation.*

We conclude this section with the observation that there is an intriguing relationship between the notion of letter graphs and geometric grid classes of permutations. We discuss this relationship in the next section.

8.3 Permutations and Graphs

From Section 2.2.3 we know that with every permutation π one can uniquely associate a graph G_π, called the *permutation graph* of π. We observe, however, that the mapping $f : \pi \to G_\pi$ is generally neither injective nor surjective. It is not injective, because different permutations can be mapped by f to the same (up to isomorphism) graph. To give an example, consider two permutations, 2413 and 3142. It is not difficult to see that the permutation graph of both of them is a path on four vertices P_4. The mapping f is not surjective, because there are graphs that are not permutation graphs, as we have seen in Section 2.2.3.

In spite of the fact that the correspondence between permutations and permutation graphs is not bijective, it allows us to establish or reveal many interesting and important connections between the two areas. It is also a powerful tool for transferring results between the areas and a fruitful source for obtaining new results. In the present section, we discuss some of the many connections between the theory of permutations and that of graphs and describe several results that connect the two areas. We start by defining some terminology.

Increasing and decreasing subsequences in permutations versus independent sets and cliques in graphs. It is not difficult to see that an increasing subsequence of elements of a permutation π corresponds to an independent set in the permutation graph G_π of π. Similarly, a decreasing subsequence in π corresponds to a clique in G_π. According to Ramsey's Theorem, every "sufficiently large" graph contains either a "big" clique or a "big" independent set. This immediately implies that a "sufficiently large" permutation contains either a "long" increasing subsequence or a "long" decreasing subsequence. Finding Ramsey numbers for graphs, i.e. determining how large the graph should be to guarantee the presence of a clique or an independent set of a given size k, is a very difficult question and this question is open even for $k = 5$.

> Joel Spencer tells a story about Erdős asking one to imagine an alien force, vastly more powerful than us, landing on Earth and demanding the value of $R(5,5)$ or they will destroy our planet. In that case, he claims, we should

> marshal all our computers and all our mathematicians and attempt to find the value. But suppose, instead, that they ask for $R(6,6)$. In that case, he believes, we should attempt to destroy the aliens.

For permutations, and hence for permutation graphs, the corresponding question is simpler and the answer was given by Erdős and Szekeres in [53]: every permutation of length k^2+1 contains either a decreasing or increasing subsequence of length $k+1$.

Patterns in permutations versus induced subgraphs in graphs. The examples of increasing and decreasing subsequences in a permutation lead naturally to the more general notion of a *pattern* in a permutation, which is analogous to the notion of a pattern in a word (see Definition 7.7.1).

Definition 8.3.1. We say that a permutation π *contains* (or *involves*) a permutation σ as a pattern if π contains a subsequence of length $|\sigma|$ whose elements appear in π in the same relative order as the elements of σ. If π does not contain σ as a pattern, then we say that π *avoids* σ.

Example 8.3.2. Consider the permutation $\pi = 261435$. It contains the increasing subsequence 245, which is equivalent to saying that π contains the permutation 123 as a pattern. Similarly, π contains the permutation 321 as a pattern, because it contains the decreasing subsequence 643. Moreover, π contains the permutation 2413 as a pattern, because it contains the subsequence 2615 and in both of them (in the permutation 2413 and in the subsequence 2615) the elements appear in the following relative order: the second smallest, the largest, the smallest and the second largest.

The notion of a pattern defines a partial order on the set of permutations known as the *pattern containment* (or *pattern involvement*) relation. It is not difficult to see that this partial order corresponds to the induced subgraph relation on permutation graphs, which is recorded in the following proposition.

Proposition 8.3.3. *If a permutation π contains a permutation σ as a pattern, then the graph G_π contains the graph G_σ as an induced subgraph.*

We observe that the converse statement is generally not true, as different permutations may correspond to the same (up to isomorphism) permutation graph.

Classes of permutations versus classes of graphs. Similarly to hereditary classes of graphs, i.e. classes that are downward closed under the induced subgraph relation, one can think of "hereditary" classes of permutations, i.e. those that are downward closed under the pattern containment relation. Such classes are known as *pattern classes* of permutations, or simply permutation classes. Unlike the area of hereditary classes of graphs, which has seen decades of study, thousands of research articles and dozens of monographs, the area of permutation classes is relatively new. It was born independently of the research on classes of graphs with its own terminology, results and research agenda. Below we compare some terminology related to classes of graphs and permutations.

Sometimes, pattern classes of permutations are called *pattern avoiding* to emphasize the fact that permutation classes can be described, similarly to hereditary classes of graphs, in terms of obstructions, i.e. the elements that do not belong to the class. For hereditary classes, such obstructions are known as *forbidden induced subgraphs.* For permutation classes, such obstructions are called *restrictions.* What are known as *minimal forbidden induced subgraphs* in the terminology of graphs correspond to *basic restrictions* in the terminology of permutations. The *set of minimal forbidden induced subgraphs* for a hereditary class corresponds to the *basis* of a permutation class, and a *finitely defined class of graphs* (i.e. a class defined by finitely many forbidden induced subgraphs) corresponds to a *finitely based permutation class* (i.e. a class whose basis consists of finitely many permutations).

Intervals in permutations versus modules in graphs. In a permutation, an *interval* is a factor (i.e. a subsequence of contiguous elements) that contains a set of contiguous values. Consider, for instance, the permutation $\pi = 261435$. The subsequence 435 is a factor of π (their elements appear contiguously in π) and its values form a contiguous set, therefore, 435 is an interval in π. The subsequence 1435 is also a factor, but its values are not contiguous (value 2 is missing), and hence this is not an interval.

Recall from Definition 5.4.5 that a module in a graph G is a subset of vertices indistinguishable by the vertices outside of the subset. This is a notion in graph theory which corresponds to intervals in permutations. Given a graph G, a subset $U \subset V(G)$ and a vertex $v \notin U$, we say that v *distinguishes* U if it has both a neighbour and a non-neighbour in U. With some care, the reader can derive the following conclusion from the definitions.

Proposition 8.3.4. *A set of elements of a permutation π forms an interval if and only if the corresponding vertices of the permutation graph G_π form a module.*

Clearly, every vertex of a graph forms a module, and the set of all vertices of the graph forms a module. These modules are called *trivial.* A graph, each module of which is trivial, is *prime.* In the theory of permutations, the notion corresponding to prime graphs is known as *simple permutations.* The following proposition is not difficult to prove.

Proposition 8.3.5. *A permutation π is simple if and only if G_π is prime.*

The notions of prime graphs and simple permutations are very helpful for various problems on graphs and permutations. In particular, the question of deciding whether a class of graphs is well-quasi-ordered by the induced subgraph relation can be reduced to prime graphs in the class [100]. To illustrate this idea, consider the class of P_4-free graphs. In [41], it was shown that every graph in this class with at least two vertices is either disconnected or the complement of a disconnected graph. Therefore, every graph with at least three vertices in this class contains a non-trivial module and hence is not prime. By the results in [100], this immediately implies that the class of P_4-free graphs is well-quasi-ordered by the induced subgraph relation. For the first time, this conclusion was derived in [45] without referring to the notion of prime graphs.

Now let us observe that the class of P_4-free graphs is a subclass of permutation graphs. Earlier we have seen that there are two permutations whose permutation graph is a P_4: these are 2413 and 3142. Moreover, it is not difficult to see that these are the *only* permutations whose permutation graph is a P_4. Therefore, we have the following statement.

Proposition 8.3.6. *A permutation* π *avoids both* 2413 *and* 3142 *if and only if* G_π *is* P_4*-free.*

It may seem that this proposition together with the well-quasi-orderability of P_4-free graphs imply well-quasi-orderability of the permutations avoiding both 2413 and 3142. Strictly speaking, this implication is not correct, as non-comparable permutations (with respect to the pattern containment relation) may correspond to comparable permutation graphs (with respect to the induced subgraph relation). However, in this particular case, both sets (of graphs and permutations) are well-quasi-ordered, and the well-quasi-orderability of permutations avoiding 2413 and 3142 was shown in [12]. This result can be viewed as a generalization of the well-quasi-orderability of P_4-free graphs, because an infinite antichain of permutation graphs necessarily implies an infinite antichain of permutations. In the next section, we discuss this relationship between graphs and permutations in more detail.

8.3.1 Well-quasi-ordered Sets of Permutations and Graphs

The relationship between well-quasi-ordered sets of permutations and graphs can be characterized by the following proposition, which follows directly from Proposition 8.3.3.

Proposition 8.3.7. *If a permutation class* Π *is well-quasi-ordered by the pattern containment relation, then the corresponding class* $\mathcal{G}_\Pi$ *of permutation graphs is well-quasi-ordered by the induced subgraph relation. Alternatively, if* $\mathcal{G}_\Pi$ *contains an infinite antichain, then so does* Π.

This proposition implies that any infinite antichain of permutation graphs gives rise to an infinite antichain of permutations. Let us observe that the question of the existence of an infinite antichain of permutations is not so obvious and the only purpose of [127] was to exhibit such an example. On the contrary, for graphs the existence of an infinite antichain with respect to the induced subgraph relation is a simple question and a trivial example is given by cycles. Unfortunately, cycles are not permutation graphs, except for C_3 and C_4. However, there is one more simple example, known as H-graphs. These are graphs of the form H_k represented in Figure 8.2.

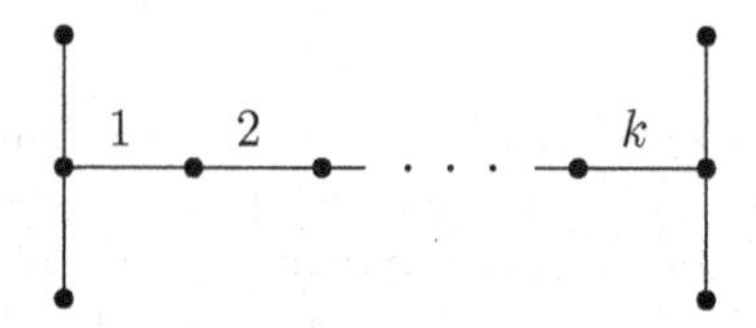

Figure 8.2: The graph H_k

It is easy to see that any two H-graphs H_i and H_j with $i \neq j$ are incomparable with respect to the induced subgraph relation. Therefore, the set of all H-graphs forms an infinite antichain with respect to this relation. Also, it is not difficult to see that the H-graphs are permutation graphs, and therefore, the corresponding permutations form an infinite antichain with respect to the pattern containment relation. Moreover, as we know from Chapter 2, the complements of permutation graphs are permutation graphs too. Hence, the permutations corresponding to the complements of H-graphs also form an infinite antichain with respect to the pattern containment relation, and this is precisely the antichain presented in [127].

In what follows we describe a more interesting example with a deeper interrelation between permutations and graphs.

8.3.2 Grid Permutation Classes and k-Letter Permutation Graphs

We begin with some definitions from [4]. Suppose that M is a $0/\pm1$ matrix. The *standard figure* of M is the set of points in $\mathbb{R}^2$ consisting of

- the increasing open line segment from $(k-1, \ell-1)$ to (k, ℓ) if $M_{k,\ell} = 1$ or
- the decreasing open line segment from $(k-1, \ell)$ to $(k, \ell-1)$ if $M_{k,\ell} = -1$.

We index matrices first by column, counting left to right, and then by row, counting bottom to top throughout. The *geometric grid class* of M, denoted by $\mathrm{Geom}(M)$, is then the set of all permutations that can be drawn on this figure in the following manner. Choose n points in the figure, no two on a common horizontal or vertical line. Then label the points from 1 to n from bottom to top and record these labels reading left to right. Figure 8.3 represents two permutations that lie, respectively, in the grid classes of

$$\begin{pmatrix} 1 & -1 \\ -1 & 1 \end{pmatrix} \text{ and } \begin{pmatrix} -1 & 1 \\ 1 & -1 \end{pmatrix}.$$

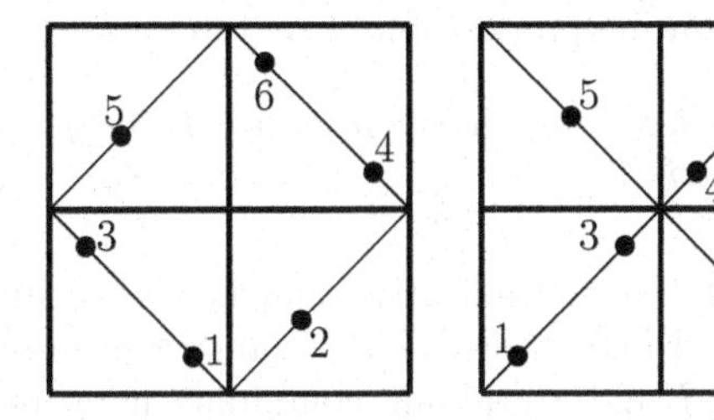

Figure 8.3: The permutation 351624 on the left and the permutation 153426 on the right

A permutation class is said to be *geometrically griddable* if it is contained in some geometric grid class. The geometrically griddable classes of permutations enjoy many nice properties. In particular, in [4] the following results have been proved.

Theorem 8.3.8. *([4]) Every geometrically griddable class of permutations is well-quasi-ordered and is in bijection with a regular language.*

In what follows, we will reveal an intriguing relationship between geometrically griddable classes of permutations and k-letter graphs, which can be stated as follows.

Theorem 8.3.9. *Let X be a class of permutations and $\mathcal{G}_X$ the corresponding class of permutation graphs. Then X is geometrically griddable if and only if $\mathcal{G}_X$ is a class of k-letter graphs for a finite value of k.*

To prove Theorem 8.3.9, we first outline the correspondence (bijection) between a geometrically griddable class and a regular language established in Theorem 8.3.8. To this end, we need the following definition from [4].

Definition 8.3.10. We say that a $0/\pm 1$ matrix M of size $t \times u$ is a *partial multiplication matrix* if there are column and row signs

$$c_1, \ldots, c_t, r_1, \ldots, r_u \in \{1, -1\}$$

such that every entry $M_{k,\ell}$ is equal to either 0 or the product $c_k r_\ell$.

Example 8.3.11. The matrix $\begin{pmatrix} 1 & 0 & -1 \\ -1 & 1 & 0 \end{pmatrix}$ is a partial multiplication matrix. This matrix has column and row signs $c_2 = c_3 = r_1 = 1$ and $c_1 = r_2 = -1$.

The importance of this notion for the study of geometric grid classes of permutations is due to the following proposition proved in [4].

Proposition 8.3.12. *([4]) Every geometric grid class is the geometric grid class of a partial multiplication matrix.*

Let M be a $t \times u$ partial multiplication matrix with column and row signs $c_1, \ldots, c_t, r_1, \ldots, r_u \in \{1, -1\}$ and let Φ_M be the *standard gridded figure* of M, that is, the standard figure of M and a gridding containing it in its interior. We will interpret the signs of the columns and rows of M as the "directions" associated with the columns and rows of Φ with the following convention: $c_i = 1$ corresponds to $\rightarrow$, $c_i = -1$ corresponds to $\leftarrow$, $r_i = 1$ corresponds to $\uparrow$, and $r_i = -1$ corresponds to $\downarrow$. The standard gridded figure of the matrix $M = \begin{pmatrix} 0 & 1 & 1 \\ 1 & -1 & -1 \end{pmatrix}$ with row signs $r_1 = -1$ and $r_2 = 1$ and column signs $c_1 = -1$, $c_2 = c_3 = 1$ is represented in Figure 8.4.

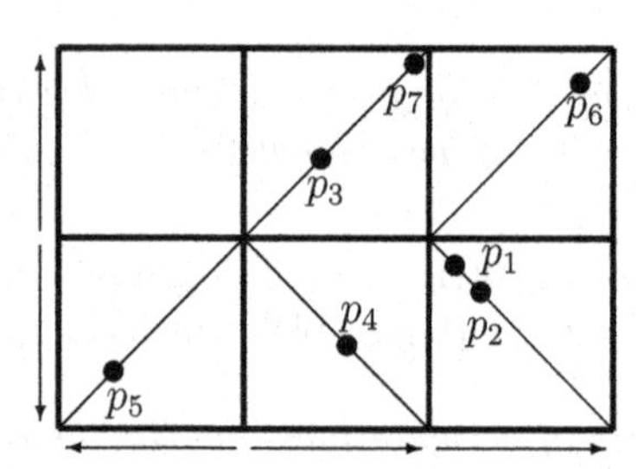

Figure 8.4: A standard gridded figure of a partial multiplication matrix

The *base point* of a cell $C_{k,\ell}$ of the figure Φ is that corner of the cell where both directions (associated with column k and row ℓ) start. For instance, in Figure 8.4 the base point of the cell $C_{3,1}$ is the top left corner.

In order to establish a bijection between Geom(M) and a regular language, we first fix an alphabet Σ (known as the *cell alphabet* of M) as follows:

$$\Sigma = \{a_{k\ell} \ : \ M_{k,\ell} \neq 0\}.$$

Now, let π be a permutation in Geom(M), i.e. a permutation represented by a set of n points in the figure Φ_M. For each point p_i of π, let d_i be the distance from the base point of the cell containing p_i to p_i. Without loss of generality, we assume that these distances are pairwise different and the points are ordered so that $0 < d_1 < d_2 < \cdots < d_n < 1$. If p_i belongs to the cell $C_{k,\ell}$ of Φ_M, we define $\phi(p_i) = a_{k\ell}$. Then $\phi(\pi) = \phi(p_1)\phi(p_2)\cdots\phi(p_n)$ is a word in the alphabet, i.e. ϕ defines a mapping from Geom(M) to Σ^*. Figure 8.4 shows seven points defining the permutation 1527436. The mapping ϕ associates with this permutation a word in the alphabet Σ as follows: $\phi(1527436) = a_{31}a_{31}a_{22}a_{21}a_{11}a_{32}a_{22}$.

Conversely, let $w = w_1 \cdots w_n$ be a word in Σ^* and let $0 < d_1 < \cdots < d_n < 1$ be n distances chosen arbitrarily. If $w_i = a_{k\ell}$, we let p_i be the point on the line segment in cell $C_{k,\ell}$ at distance d_i from the base point of $C_{k,\ell}$. The n points of Φ_M constructed in this way define a permutation $\psi(w)$ in Geom(M). Therefore, ψ is a mapping from Σ^* to Geom(M).

This correspondence between Σ^* and Geom(M) is not yet a bijection, as illustrated in Figure 8.5, because the order in which the points are consecutively inserted into *independent* cells (i.e. cells which share neither a column nor a row) is irrelevant. To turn this correspondence into a bijection, we say that two words $v, w \in \Sigma^*$ are *equivalent* if one can be obtained from the other by successively interchanging adjacent letters which represent independent cells. The equivalence classes of this relation form a *trace monoid* and each element of this monoid is called a *trace*. It is known that in any trace monoid it is possible to choose a unique representative from each trace in such a way that the resulting set of representatives forms a regular language. This is the language which is in a bijection with Geom(M), as was shown in [4].

Next, we will show that the permutation graph G_π of $\pi \in$ Geom(M) is a k-letter graph with $k = |\Sigma|$. Indeed, the set of non-empty cells of the figure Φ_M defines a partition of the vertex set of G_π into cliques and independent sets and the word $\phi(\pi)$ defines the order of the vertex set of G_π satisfying the conditions of Theorem 8.2.2. More formally, let us show that the matrix M uniquely defines a set $\mathcal{P} \subseteq \Sigma^2$ such that the letter graph $G(\mathcal{P}, w)$ of the word $w = \phi(\pi)$ coincides with G_π. In order to define the set $\mathcal{P}$, we observe that two points p_i and p_j of a permutation $\pi \in$ Geom(M) correspond to a pair of adjacent vertices in G_π if and only if one of them lies to the left and above the second one in the figure Φ_M. Therefore, if

- $M_{k,\ell} = 1$, then the points lying in the cell $C_{k,\ell}$ form an independent set in the

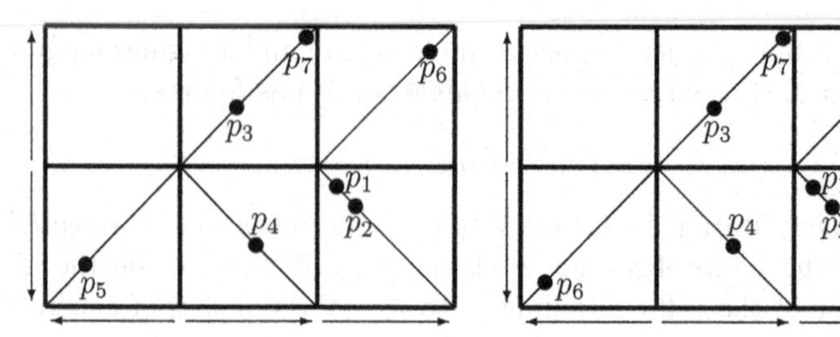

Figure 8.5: Two drawings of the permutation 1527436. The drawing on the left is encoded as $a_{31}a_{31}a_{22}a_{21}a_{11}a_{32}a_{22}$ and the drawing on the right is encoded as $a_{31}a_{31}a_{22}a_{21}a_{32}a_{11}a_{22}$

permutation graph of π. Therefore, we do not include the pair $(a_{k\ell}, a_{k\ell})$ in $\mathcal{P}$.

- $M_{k,\ell} = -1$, then the points lying in the cell $C_{k,\ell}$ form a clique in the permutation graph of π. Therefore, we include the pair $(a_{k\ell}, a_{k\ell})$ in $\mathcal{P}$.
- two cells $C_{k,\ell}$ and $C_{s,t}$ are independent with $k < s$ and $\ell < t$, then no point of $C_{k,\ell}$ is adjacent to any point of $C_{s,t}$ in the permutation graph of π. Therefore, we include neither $(a_{k\ell}, a_{st})$ nor $(a_{st}, a_{k\ell})$ in $\mathcal{P}$.
- two cells $C_{k,\ell}$ and $C_{s,t}$ are independent with $k < s$ and $\ell > t$, then every point of $C_{k,\ell}$ is adjacent to every point of $C_{s,t}$ in the permutation graph of π. Therefore, we include both pairs $(a_{k\ell}, a_{st})$ and $(a_{st}, a_{k\ell})$ in $\mathcal{P}$.
- two cells $C_{k,\ell}$ and $C_{s,t}$ share a column, i.e. $k = s$, then we look at the sign (direction) associated with this column and the relative position of the two cells within the column.
  - If $c_k = 1$ (i.e. the column is oriented from left to right) and $\ell > t$ (the first of the two cells is above the second one), then only the pair $(a_{k\ell}, a_{kt})$ is included in $\mathcal{P}$.
  - If $c_k = 1$ and $\ell < t$, then only the pair $(a_{kt}, a_{k\ell})$ is included in $\mathcal{P}$.
  - If $c_k = -1$ (i.e. the column is oriented from right to left) and $\ell > t$ (the first of the two cells is above the second one), then only the pair $(a_{kt}, a_{k\ell})$ is included in $\mathcal{P}$.
  - If $c_k = -1$ and $\ell < t$, then only the pair $(a_{k\ell}, a_{kt})$ is included in $\mathcal{P}$.
- two cells $C_{k,\ell}$ and $C_{s,t}$ share a row, i.e. $\ell = t$, then we look at the sign (direction) associated with this row and the relative position of the two cells within the row.

- If $r_\ell = 1$ (i.e. the row is oriented from bottom to top) and $k < s$ (the first of the two cells is to the left of the second one), then only the pair $(a_{k\ell}, a_{s\ell})$ is included in $\mathcal{P}$.
- If $r_\ell = 1$ and $k > s$, then only the pair $(a_{s\ell}, a_{k\ell})$ is included in $\mathcal{P}$.
- If $r_\ell = -1$ (i.e. the row is oriented from top to bottom) and $k < s$, then only the pair $(a_{s\ell}, a_{k\ell})$ is included in $\mathcal{P}$.
- If $r_\ell = -1$ and $k > s$, then only the pair $(a_{k\ell}, a_{s\ell})$ is included in $\mathcal{P}$.

It is now a routine task to verify that $G(\mathcal{P}, w)$ coincides with G_π.

8.3.3 Universal Graphs and Universal Permutations

Let X be a family of graphs and X_n the set of n-vertex graphs in X. A graph containing all graphs from X_n as induced subgraphs is called *n-universal* for X. The problem of constructing universal graphs is closely related to graph representations and finds applications in theoretical computer science [85]. This problem is trivial if universality is the only requirement, since the union of all vertex disjoint graphs from X_n is obviously n-universal for X. However, this construction is generally neither optimal, in terms of the number of its vertices, nor proper, in the sense that it does not necessarily belong to X.

Let us denote an n-universal graph for X by $U_X^{(n)}$ and the set of its vertices by $V(U_X^{(n)})$. Since the number of n-vertex subsets of $V(U_X^{(n)})$ cannot be smaller than the number of graphs in X_n, we conclude that

$$\log_2 |X_n| \leq \log_2 \binom{|V(U^{(n)})|}{n} \leq n \log_2 |V(U^{(n)})|.$$

Also, trivially, $n \leq |V(U_X^{(n)})|$, and hence,

$$n \log_2 n \leq n \log_2 |V(U^{(n)})|.$$

We say that $U_X^{(n)}$ is *optimal* and *asymptotically optimal*, respectively, if $n \log_2 |V(U^{(n)})| = \max(\log_2 |X_n|, n \log_2 n)$ and

$$\lim_{n\to\infty} \frac{n \log_2 |V(U^{(n)})|}{\max(\log_2 |X_n|, n \log_2 n)} = 1.$$

Also, $U_X^{(n)}$ is *order-optimal* if there is a constant c such that for all $n \geq 1$,

$$\frac{n \log_2 |V(U^{(n)})|}{\max(\log_2 |X_n|, n \log_2 n)} \leq c.$$

Optimal (asymptotically optimal, order-optimal) n-universal graphs have been constructed for many families of graphs such as planar graphs [38], graphs of bounded arboricity [11], of bounded vertex degree [31, 55] etc.

We say that $U_X^{(n)}$ is a *proper* n-universal graph for X if it belongs to X. Proper optimal (asymptotically optimal, order-optimal) n-universal graphs have been constructed for the class of all graphs [111], threshold graphs [76], split graphs, bipartite graphs [106] and bipartite permutation graphs [107].

The class of bipartite permutation graphs is precisely the class of triangle-free permutation graphs, because bipartite graphs are graphs without odd cycles and odd cycles of length at least 5 are not permutation graphs. Clearly 321 is the only permutation whose permutation graph is a triangle. Therefore, the class of triangle-free permutation graphs corresponds to 321-avoiding permutations (see Definition 8.3.1), i.e. a permutation π is 321-avoiding if and only if its permutation graph is triangle-free. This correspondence between bipartite permutation graphs and 321-avoiding permutations allowed the authors of [14] to transform the universal construction for bipartite permutation graphs into a universal construction for 321-avoiding permutations. Both constructions use n^2 elements, i.e. there is a bipartite permutation graph with n^2 vertices containing all bipartite permutation graphs with n vertices as induced subgraphs and there is a 321-avoiding permutation of length n^2 containing all permutations of length n as patterns. The paper [14] also reveals an interesting relationship between 321-avoiding permutations and split permutation graphs, which allows us to transform the universal construction for 321-avoiding permutations into a proper n-universal split permutation graph with $4n^3$ vertices.

In the permutation literature, universal permutations are also known as *superpatterns* [54, 110]. In the case of all permutations of length n, the current best upper bound for the size of a superpattern is $n^2/2 + \Theta(n)$ [110]. Superpatterns for specific families of permutations were introduced in [19, 20]. In particular, [20] improves the n^2 universal construction for 321-avoiding permutations to a superpattern of size $O(n^{3/2})$, while [19] describes 213-avoiding superpatterns of size $n^2/4 + \Theta(n)$. The authors also show that every proper subclass of the 213-avoiding permutations has near-linear superpatterns. Small superpatterns described in [19] and [20] were used to construct so-called *universal point sets* for graph drawing. These results reveal one more connection between graphs and permutations.

8.4 Almost Periodic Words and a Jump to the Bell Number for Hereditary Graph Properties

We recall from Chapter 2 that the speed of a hereditary property X is the number of n-vertex *labelled* graphs in X.

The speeds of hereditary properties and their asymptotic structures have been extensively studied, originally in the special case of a single forbidden subgraph [51, 52, 98, 117, 118, 119], and more recently in the general case [6, 9, 16, 17, 18, 123]. These studies showed, in particular, that there is a certain correlation between the speed of a property X and the structure of graphs in X, and that the rates of the speed growth constitute discrete layers. The lowest four layers have been distinguished in [123]: these are constant, polynomial, exponential and factorial. In other words, the authors of [123] showed that some classes of functions do not appear as the speed of any hereditary property, and that there are discrete jumps, for example, from polynomial to exponential speeds.

Independently, similar results were obtained by Alekseev in [7]. Moreover, Alekseev provided the first four layers with the description of all minimal classes, that is, he identified in each layer a family of classes, every hereditary subclass of which belongs to a lower layer (see also [16] for some more involved results). In each of the lowest four layers the set of minimal classes is finite and each of them is defined by finitely many forbidden induced subgraphs. This provides an efficient way of determining whether a property X belongs to one of the first three layers.

One more jump in the speed of hereditary properties was identified in [18] and it separates the properties with speeds strictly below the Bell number B_n (see Definition B.2.4) from those whose speed is at least B_n. With a slight abuse of terminology, we will refer to these two families of graph properties as properties below and above the Bell number, respectively. The importance of this jump is due to the fact that all the properties below the Bell number are well structured. In particular, all of them have bounded clique-width [8] and all of them are well-quasi-ordered by the induced subgraph relation [100]. From the results in [100] it follows that all hereditary properties below the Bell number can be characterized by finitely many forbidden induced subgraphs and hence each of them can be recognized in polynomial time.

In spite of the importance of the jump to the Bell number, until recently very little was known about the boundary separating the two families, that is, very little was known about the *minimal* classes above the Bell number. The paper [18] distinguishes two cases in the study of this question: the case where a certain parameter associated with each class of graphs, called in [13] the *distinguishing number*, is finite and the case where this parameter is infinite. For the case where the distinguishing number is infinite, the paper [18] provides a complete description of minimal classes, of which there are precisely 13. In the case where this parameter is finite, the family of minimal classes is infinite and all of them have been characterized in [13] via the notion of almost periodic words as follows.

Let A be a finite alphabet, w a word over A, and H an undirected graph with loops allowed and with vertex set $V(H) = A$. For any increasing sequence $u_1 < \cdots < u_m$ of positive integers define $G_{w,H}(u_1, \ldots, u_m)$ to be the graph with vertex set $\{u_1, \ldots, u_m\}$ and an edge between u_i and u_j if and only if

- either $|u_i - u_j| = 1$ and $w_{u_i}w_{u_j} \notin E(H)$,
- or $|u_i - u_j| > 1$ and $w_{u_i}w_{u_j} \in E(H)$.

Define $\mathcal{P}(w, H)$ to be the hereditary class consisting of graphs $G_{w,H}(u_1, \ldots, u_m)$ for all finite increasing sequences $u_1 < \cdots < u_m$ of positive integers.

Recall the definition of a factor in a word in Definition 3.2.8.

Definition 8.4.1. A word w is called *almost periodic* if for any factor f of w there is a constant k_f such that any factor of w of size at least k_f contains f as a factor.

Theorem 8.4.2. *([13]) Let X be a hereditary class of graphs with a finite distinguishing number. Then X is a minimal class above the Bell number if and only if there exists a finite graph H and an infinite almost periodic word w over $V(H)$ such that $X = \mathcal{P}(w, H)$.*

8.5 From Monomials to Words and to Graphs

The content of this section is based on the results from [58]. To state the results we need to extend some terminology of partially ordered sets introduced in Section 8.2.

> **Detour:** *lower and upper sets.*
>
> Let $(A, \leq)$ be a partially ordered set (poset), i.e. a set A with a partial order $\leq$ on A. A subset $B \subseteq A$ is
>
> – a *lower set* (also known as a *down set*, a *downward closed set*, a *lower ideal* or simply an *ideal*) if $x \in B$ implies $y \in B$ for every $y \in A$ such that $y \leq x$.

– an *upper set* (also known as an *upward closed set* or an *upper ideal*) if $x \in B$ implies $y \in B$ for every $y \in A$ such that each $y \geq x$.

We observe that in [58] the word "ideal" is used for upper sets.

Clearly, $B \subseteq A$ is a lower set if and only if $A - B$ is an upper set. Following [58], for an arbitrary set $M \subseteq A$, we denote by $\langle M \rangle$ the upper set generated by M, i.e. the set containing all elements of M and all elements of A above at least one element of M. It is well known (and not difficult to see) that if A is well founded, then any upper set B in A is generated by a unique (not necessarily finite) set of its minimal elements, called the *minimal generating set* of B. If the minimal generating set of B is finite, B is called *finitely generated.* The minimal generating set of an upper set B is also the unique set of minimal elements not in $A - B$, i.e. the set of minimal forbidden elements for the lower set $A - B$. Due to the minimality of the elements in the minimal generating set, they form an antichain.

Let $X = \{x_1, \ldots, x_n\}$ be a finite alphabet, X^* the *free monoid* on X (the set of all finite words over X) and $[X]$ the *free commutative monoid* on X (see Definition B.1.13). We call the members of $[X]$ *monomials* and represent them multiplicatively as $x_1^{i_1} \cdots x_n^{i_n}$. In this section, we consider the factor containment relation on words and the divisibility relation on monomials. Recall Definition 3.2.8 saying that a word w is a factor of the word u if w can be obtained from u by deleting some (possibly empty) prefix and some (possibly empty) suffix. Also, we say that a monomial $w = x_1^{i_1} \cdots x_n^{i_n}$ *divides* a monomial $u = x_1^{j_1} \cdots x_n^{j_n}$ if $i_1 \leq j_1, \ldots, i_n \leq j_n$. Clearly, both relations are well-founded partial orders.

Following [58], we denote by π the *canonical monoid epimorphism* $X^* \to [X]$. It maps a word w in X^* to the unique monomial indicating how many times each letter of X appears in w. For example, $\pi(x_1x_2x_2x_1x_3) = x_1^2x_2^2x_3$. This mapping is not injective, i.e. different words can be mapped to the same monomial. To define an inverse mapping, we fix an arbitrary linear order $\prec$ on the letters of the alphabet and call a word w in X^* *sorted* if its letters appear in non-decreasing order (with respect to $\prec$). Let $\sigma_\prec$ be the function that maps each monomial $u \in [X]$ to the unique sorted word in $\pi^{-1}(u)$, i.e. $\pi\sigma_\prec$ is the identity map on $[X]$.

It follows from Dickson's Lemma (also from Higman's Lemma) that the divisibility relation on monomials contains no infinite antichains, and therefore, any upper set I in $[X]$ is finitely generated. But what about $\langle \sigma_\prec(I) \rangle$, i.e. the upper set in X^* generated by the sorted words corresponding to the elements of I? Is it finitely generated? This question was studied in [58], and was answered there as follows. Given an ordering $\prec$ of X, we say that a letter x is *extremal* in a monomial $v \in [X]$ if x is the smallest or the largest letter with a positive exponent in v. Every letter of X between the two extremes is said to be *internal* to v. Also, by $v \setminus x$ we denote the monomial obtained from v by erasing the letter x.

Theorem 8.5.1. *([58]) Let $I = \langle M \rangle$ be an upper set in $[X]$ generated by an antichain M of monomials and let $\prec$ be a linear order of the letters in X. Then the ideal $\langle \sigma_{\prec}(I) \rangle$ in X^* is finitely generated if and only if for every monomial $u \in M$ and every letter $x \in X$ internal to u, there is a monomial $v \in M$ such that x is extremal in v and $v \setminus x$ divides u.*

This theorem shows that algorithmically the above question is simple if the upper set I is given by its (finite) generating set M. A more difficult question is whether there *exists* an ordering $\prec$ such that $\langle \sigma_{\prec}(I) \rangle$ is finitely generated. In [58], an ordering of X satisfying the conditions of Theorem 8.5.1 was called *cool* for an antichain of monomials M. Also, the authors of [58] proved that the problem of determining whether there exists a cool ordering for M

- is NP-complete even if M consists of quadratic monomials only, i.e. monomials of the form $x_i x_j$ or x_i^2;
- can be solved in polynomial time if M is square-free, i.e. if every monomial contains only letters of degree at most 1.

Both results were obtained by means of graph theory. To this end, it was first observed that the square-free case can be reduced to the quadratic case, because if M is a square-free antichain that admits a cool ordering then it contains no monomial of total degree more than 2. Degree 1 monomials are trivially handled, so the square-free sets M of interest consist of quadratic monomials only.

To handle the quadratic monomials, the authors of [58] describe M by a graph $G(M)$ as follows. The vertex set of $G(M)$ is X and two distinct vertices x_i and x_j are adjacent if and only if $x_i x_j$ is *not* in M. Also, let T_M denote the set of letters (vertices) whose square is not in M.

An orientation of a graph is said to be *transitive at a vertex y* if, whenever oriented edges $x \to y$ and $y \to z$ exist, then the edge $x \to z$ must also exist. A graph is a comparability graph if it admits an orientation that is transitive at all its vertices; such an orientation is always acyclic.

Theorem 8.5.2. *([58]) A set M of quadratic monomials admits a cool order if and only if the graph $G(M)$ admits an acyclic orientation that is transitive at all vertices of T_M. In particular, if M is square-free, it admits a cool order if and only if $G(M)$ is a comparability graph.*

Since recognition of comparability graphs can be done in polynomial time, the above theorem implies that deciding whether a square-free set of monomials admits a cool ordering is a polynomially solvable task. For the NP-completeness of this task in the non-square-free case, the authors of [58] proved the following result.

Theorem 8.5.3. *([58]) The problem:*

> *Given a graph G and a set $T \subsetneq V(G)$, is there an acyclic orientation of G that is transitive at all vertices in T?*

is NP-complete.

8.6 Polygon-Circle Graphs

Definition 8.6.1. A *polygon-circle graph* is the intersection graph of a set of polygons inscribed in a circle. That is, two vertices of a graph are connected by an edge if the respective polygons have a non-empty intersection.

Example 8.6.2. See Figure 8.6 for an example of a polygon-circle graph on three vertices.

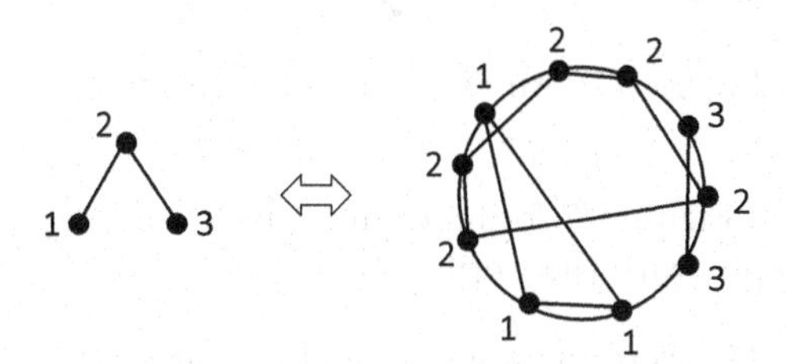

Figure 8.6: A polygon-circle graph on three vertices

Polygon-circle graphs were introduced in [97] and are also known under the name *spiders*. If G is a polygon-circle graph and each polygon in the representation of G is k sided (a k-gon), then G is a *k-polygon-circle graph*. Since a chord of a circle can be viewed as a 2-gon, polygon-circle graphs generalize circle graphs, i.e. the intersection graphs of chords of a circle. We remind the reader that the circle graphs constitute an important subclass of word-representable graphs.

If we view each polygon as a letter, and read the incidences of the polygons on the circle, say in clockwise direction, we see that two polygons intersect if and only if there *exists* a pair of occurrences of the two letters that alternate. For example, reading the letters in the right picture in Figure 8.6 starting from the top 3, we obtain the word 3231122122, which defines the edges $(1,2)$ and $(2,3)$. Thus, we again deal with representing graphs by words. However, this representation is different from the word-representable graphs case, where *all* occurrences of the two letters must alternate in order for the vertices to be adjacent.

8.7 Word-Digraphs

This section is based on [22] and [23] and it deals with *word-digraphs*, called *word-graphs* in the original sources, which are directed graphs of a special type.

Definition 8.7.1. A word-digraph $G_w = (V(G_w), E(G_w))$ is a simple digraph associated with a word $w = w_1 w_2 \cdots w_\ell$ over an n-letter alphabet for $\ell > n$ such that the vertex set $V(G_w) = A(w)$, the alphabet of w, and the directed edges are defined by

$$E(G_w) = \{(w_1, w_2), (w_2, w_3), \ldots, (w_{\ell-1}, w_\ell)\},$$

where $w_i \neq w_{i+1}$ for $1 \leq i \leq \ell - 1$.

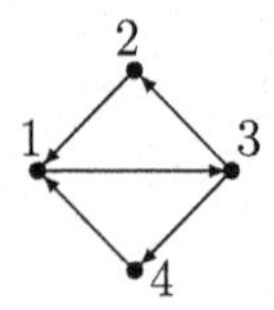

Figure 8.7: An example of a word-digraph

Example 8.7.2. See Figure 8.7 for an example of a word-digraph. This digraph is associated, for example, with the word 322134441.

According to [23], word-digraphs have close links with a range of stochastic word-based processes such as language, music, DNA and protein sequences. Word-digraphs are also closely related to the notion of digraph path decomposition [10].

Figure 8.8: Non-word-digraphs on three vertices

Not all digraphs are word-digraphs: see Figure 8.8 for all non-word-digraphs on three vertices, where a bidirectional edge is a way to represent two edges with opposite orientations. Next, we present a number of selected results on word-digraphs starting with a characterization result.

Theorem 8.7.3. *([22, 23]) A digraph D is a word-digraph if and only if there exists a directed path P in D such that P transverses each edge of D at least once. The sequence of vertices visited by P determines a word w representing D.*

Example 8.7.4. For example, the word 321341 gives a directed path for the digraph in Figure 8.7 satisfying the conditions of Theorem 8.7.3.

The word w in Theorem 8.7.3 is called a *representational word* of D. As a corollary to Theorem 8.7.3 we have the following statement.

Corollary 8.7.5. *([23])*

- *Word-digraphs are weakly connected.*
- *Strongly connected digraphs are word-digraphs.*

Proof. The path P in Theorem 8.7.3 when transversing each edge at least once also transverses each vertex at least once thus connecting any pair of vertices by an undirected path, and the first statement is proved.

Due to strong connectivity, we can start a directed path at a vertex v, then reach any desired vertex u, and then come back to v. Returning to v as many times as it takes, we can traverse, by a path P, all vertices in a strongly connected graph D starting at any vertex v and coming back to v. Suppose an edge $u \to x$ is not traversed by P. We can extend P to cover this edge as follows: once u is reached in P for the first time, traverse the edge $u \to x$, then use any path to return to u (which exists due to strong connectivity) and then continue with the other steps of P. Clearly, in this manner we can traverse all the edges in D, and thus D is a word-digraph by Theorem 8.7.3. □

Remark 8.7.6. Note that our proof of the second part in Corollary 8.7.5 is similar in nature to the well known proof of the fact that a balanced and strongly connected digraph has an Eulerian cycle.

Recall Definition B.2.3 for the notion of a Stirling number of the second kind $\left\{ n \atop k \right\}$. In the next theorem, the set of words of length ℓ over an n-letter alphabet is referred to as a *word-digraph family* $\mathcal{M}(\ell, n)$. Note that by Corollary 8.7.5 every strongly connected digraph is represented by a word in some word-digraph family.

The following theorem is the main enumerative result in [23].

Theorem 8.7.7. *([23]) There exist $\varphi(\ell, n)$ strongly connected word-digraphs in $\mathcal{M}(\ell, n)$, where*

$$\varphi(\ell, n) = n!T(\ell, n) \text{ and}$$

$$T(\ell, n) = \left\{ \begin{matrix} \ell - 1 \\ n \end{matrix} \right\} + \sum_{j=0}^{\ell-2} \sum_{m=0}^{n-2} \left\{ \begin{matrix} j \\ m \end{matrix} \right\} T(\ell - j - 1, n - m)(n - m - 1).$$

A number of structural results on word-digraphs are presented in [22], of which we will mention just one. First we need the following definition.

Definition 8.7.8. Edge contraction in the case of digraphs is defined like edge contraction in the case of undirected graphs (see Definition 5.2.7) with the following exceptions. Two edges between a vertex pair with opposite orientations are treated as a single bidirectional edge. If a bidirectional edge is contracted then both edge-pairs are removed. Parallel edges of homogeneous orientation and self-loops formed by contractions are ignored so that all contractions remain simple digraphs.

Example 8.7.9. See Figure 8.9 for an example of edge contraction of a digraph.

1 2 4 3 → x 4 3

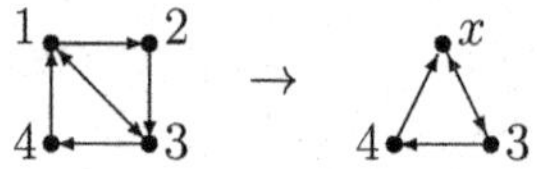

Figure 8.9: Edge contraction of a digraph. The edge $(1, 2)$ is contracted

Theorem 8.7.10. *([22]) The set of word-digraphs is closed under edge contraction.*

To illustrate Theorem 8.7.10, note that the graph to the left in Figure 8.9 can be represented by the word 1234131, while the graph to the right in this figure can be represented by the word $x34x3x$.

Remark 8.7.11. Recall from Subsection 5.4.2 that edge contraction is not a safe operation in the case of word-representable graphs. That is, starting with a word-representable graph we are not guaranteed to end up with a word-representable graph after applying edge contraction, which is not the case with word-digraphs due to Theorem 8.7.10.

8.8 Conclusion

Even though the main subject of this book is word-representable graphs, there are many other instances of interrelations between words and graphs in the literature. One well known example here is Prüfer sequences encoding trees. Less familiar examples include threshold graphs and letter graphs. The chapter also discusses permutation graphs, in particular, from a pattern avoidance point of view.

Of particular interest to us in this chapter are polygon-circle graphs, since their definition, generalizing the definition of circle graphs, can be related to alternations in words. However, while in the word-representable graph case we require strict alternation of two letters, each occurring more than once, in order for the corresponding vertices to be connected, in the case of polygon-circle graphs it is sufficient for letters to alternate at least once.

Word-digraphs, also discussed in this chapter, are an example of directed graphs representable by words.

Chapter 9

More on Interrelations Between Words and Graphs

Chapter 8 gives examples of interrelations between words and graphs in the literature. In this chapter, we provide a few more such examples that include Gray codes, the snake-in-the-box problem, de Bruijn graphs and graphs of overlapping permutations, and finding independent sets in path-schemes.

9.1 Gray Codes

The problem of generating and exhaustively listing the objects of a given class is important for several areas of science such as *computer science*, *hardware* and *software testing*, *biology* and *(bio)chemistry*.

The idea of so-called *Gray codes* (or *combinatorial Gray codes*) is to list the objects in such a way that two successive objects, encoded by words, have codes that differ as little as possible in a specified sense. In particular, a Gray code can be a list of words such that each word differs from its successor by a number of letters which is bounded independently of the length of the word.

Originally, a Gray code was used in a telegraph demonstrated by French engineer Émile Baudot in 1878. However, we say "the Gray code" to refer to the *reflected binary code* introduced by Frank Gray in 1947 to list all binary words of length n. This list can be generated recursively from the list of binary words of length $n-1$ by reflecting it (that is, listing the entries in reverse order), concatenating the original list with the reversed list, prefixing the entries in the original list with 0, and then prefixing the entries in the reflected list with 1. See Figure 9.1 for examples when $n = 1, 2, 3$.

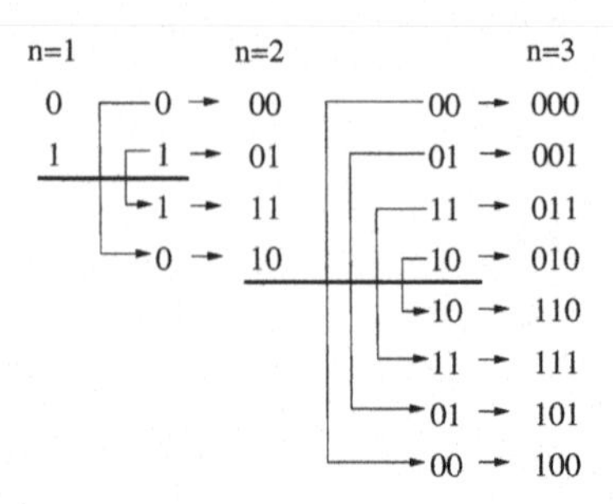

Figure 9.1: The reflected binary codes for $n = 1, 2, 3$

Definition 9.1.1. An *n-cube*, also called an *n-dimensional hypercube* or an *n-dimensional cube*, is an n-dimensional analogue of a square ($n = 2$) and a cube ($n = 3$). It is a closed, compact, convex figure whose 1-skeleton consists of groups of opposite parallel line segments aligned in each of the space's dimensions, perpendicular to each other and of the same length. Vertices of an n-cube can be labelled in a bijective way by binary strings of length n, so that adjacent vertices will differ in exactly one coordinate.

Example 9.1.2. In Figure 9.1, there are (labelled) n-cubes corresponding to the cases $n = 1, 2, 3$.

It follows from the definitions that a Gray code for binary words of length n corresponds to a Hamiltonian path (a path passing through each vertex exactly once) in the n-cube. In Figure 9.2 we show the Hamiltonian paths corresponding to the lists produced in Figure 9.1 for $n = 1, 2, 3$.

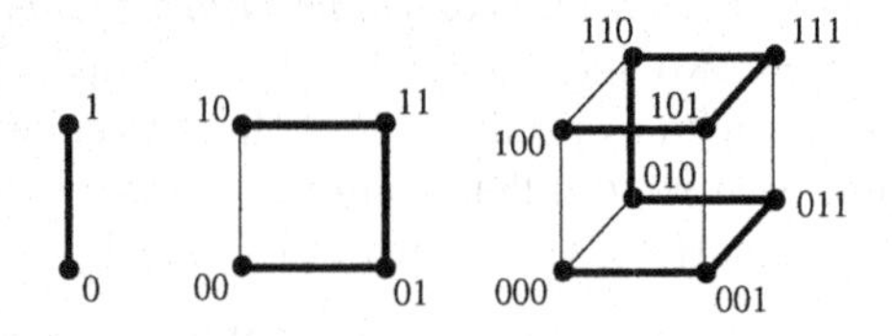

Figure 9.2: The Hamiltonian paths corresponding to the reflected binary codes for $n = 1, 2, 3$

Much has been discovered and written about the Gray code (see [48] by Doran for a survey) and it has been used, for example, in *error corrections in digital communication* and in solving the *Towers of Hanoi puzzle.*

9.2 The Snake-in-the-Box Problem

Definition 9.2.1. The *Hamming distance* between two vertices in an n-cube is the number of places in which the vertices' coordinates differ. For a vertex α in an n-cube, the vertices that have Hamming distance of 1 to α are called α's *neighbours.*

Example 9.2.2. The Hamming distance between 0001 and 1101 is 2.

The Hamming distance can be viewed as the length of the shortest path in the cube to get from one vertex to the other one.

The *snake-in-the-box problem* in graph theory and computer science deals with finding a certain kind of path along the edges of an n-cube. This path starts at one vertex and travels along the edges to as many vertices as it can reach. After it visits a new vertex, the previous vertex and all of its neighbours must be marked as unusable. The path should never travel to a vertex after it has been marked unusable. In other words, two adjacent vertices in a path have Hamming distance 1, but for non-adjacent vertices, the Hamming distance must be at least 2 (the path does not come too close to itself).

The snake-in-the-box problem was first described by Kautz [86] in 1958, motivated by the theory of *error-correcting codes.*

The maximum lengths for the snake-in-the-box problem are known for dimensions one through seven; these are

$$1, 2, 4, 7, 13, 26, 50.$$

Beyond the length $n = 7$, the exact length of the longest snake is not known; the best lengths found so far for dimensions eight through twelve are

$$98, 190, 363, 680, 1260.$$

For the snake-in-the-box problem, it is known that as n goes to infinity, the maximum length is $c2^n$ for a constant c, $0 < c < 1$. Thus, the maximum snake is rather long since the total number of vertices in an n-cube is 2^n. This result was shown by Evdokimov [56] in 1969. What makes this result special is that combinatorics on words was crucial in obtaining it; thus, combinatorics on words was helping graph theory in this case, while usually we witness the opposite situation when these two disciplines meet (e.g. in the case of *de Bruijn sequences*, *universal permutations* and *word patterns* discussed briefly in Section 9.3 below).

The rest of this section is devoted to presenting the far from trivial construction by Evdokimov [56] giving long snakes. We do not provide any proofs, nor known upper and lower bounds for the constant c studied in the literature, whose exact value is still unknown.

We begin by defining the *sequence of* σ and the *Zimin words.*

Definition 9.2.3. Any natural number n can be presented unambiguously as $n = 2^t(4s+\sigma)$, where $\sigma < 4$, and t is the largest natural number such that 2^t divides n. If n runs through the natural numbers, then σ runs through the sequence w_σ called the sequence of σ. The initial letters of w_σ are $11311331113313\cdots$.

Remark 9.2.4. The sequence w_σ is equivalent to the *Dragon curve sequence* considered in the literature.

Definition 9.2.5. The *nth Zimin word* is defined recursively as follows: $Y_1 = y_1$ and $Y_n = Y_{n-1}y_nY_{n-1}$.

Thus, Y_n is over the alphabet $\{y_1, y_2, \ldots, y_n\}$ and its length is easily seen by mathematical induction to be $2^n - 1$.

Example 9.2.6. The fourth Zimin word is $Y_4 = y_1y_2y_1y_3y_1y_2y_1y_4y_1y_2y_1y_3y_1y_2y_1$.

Zimin words appear in various contexts in the literature.

Let A and B be the prefixes of length $2^k - 1$ of the sequences $ababab\cdots$ and $cdecdecde\cdots$, respectively. Also, slightly abusing the notation, we assume now that w_σ denotes the prefix of length $2^k - 1$ of the sequence w_σ.

Further, let $X = x_1x_2\cdots x_k = (ac1)(bd1)(ae3)(bc1)\cdots$, that is, x_i can be thought of as a "super letter" formed by the ith letters of the sequences A, B and w_σ for $1 \le i \le 2^k - 1$. Clearly, some of the x_is must be the same since the alphabet of "super letters" is limited to 12 letters $(= 2\cdot 3\cdot 2)$. In fact, all 12 letters occur in X for long enough k.

Next we shuffle three sequences of length $2^k - 1$, namely X, Y_k and $Z = zz\cdots z$ as follows:

$$x_1y_1x_2y_2x_3y_1x_4y_3\cdots \text{ (shuffling } X \text{ and } Y_k\text{);}$$

$$zx_1zx_2zx_3zx_4z\cdots \text{ (shuffling } Z \text{ and } X\text{).}$$

Finally, we form the word W by shuffling the last two words:

$$W = x_1zy_1x_1x_2zy_2x_2x_3zy_1x_3\cdots.$$

The number of different letters in the word W is $n = 12 + k + 1$ corresponding to the dimension of an n-cube. The word W gives instructions on where to go in the n-cube, that is, which bit to change to make the next move. It was proved by Evdokimov [56] that W is "strongly asymmetric" and the path corresponding to it in the n-cube has desirable properties. Since the length of W is $2^{k+2} - 4 = 2^{n-11} - 4 \approx 0.0005\cdot 2^n$, we obtain the desired result.

9.3 Universal Cycles and de Bruijn Graphs

The notion of a universal object can be defined in various ways depending on the context. In Subsection 8.3.3 we discussed universal graphs and universal permutations. The basic idea here is to come up with an object, typically of smallest possible size, that contains inside it all the objects from a given class of objects. The current section not only discusses the celebrated de Bruijn graphs along with graphs of overlapping permutations, thus showing close relations between graphs and words/permutations, but also provides a sketch of a general setup over universal cycles for combinatorial structures.

9.3.1 De Bruijn Graphs and Graphs of Overlapping Permutations

Definition 9.3.1. An *n-dimensional de Bruijn* graph $\vec{D}_{n,m}$ is a directed graph on m^n vertices labelled by the words of length n over the m-letter alphabet $A_m = \{0, 1, \ldots, m-1\}$. There is a directed edge from a vertex u to a vertex v if and only if $u = aw$ and $v = wb$ for some word w over A_m of length $n-1$, and $a, b \in A_m$, that is, if the end of the word u overlaps with the beginning of v.

Example 9.3.2. See Figure 9.3 for de Bruijn graphs $\vec{D}_{1,2}$, $\vec{D}_{2,2}$ and $\vec{D}_{3,2}$.

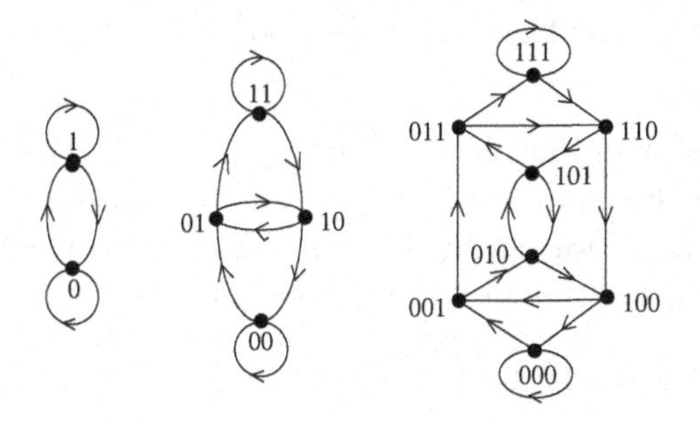

Figure 9.3: The graphs $\vec{D}_{1,2}$, $\vec{D}_{2,2}$ and $\vec{D}_{3,2}$

Theorem 9.3.3. *$\vec{D}_{n,m}$ is strongly connected and balanced, and thus by Theorem A.3.2 there exists an Eulerian cycle in $\vec{D}_{n,m}$.*

Proof. To show that $\vec{D}_{n,m}$ is strongly connected, we need to show that there is a directed path in it from any vertex $x_1x_2\cdots x_n$ to any other vertex $y_1y_2\cdots y_n$. Such a path is given by the following sequence of vertices:

$$x_1x_2\cdots x_n \to x_2x_3\cdots x_ny_1 \to x_3x_4\cdots x_ny_1y_2 \to \cdots \to y_1y_2\cdots y_n.$$

To show that $\vec{D}_{n,m}$ is balanced, we note that for any vertex $x_1x_2\cdots x_n$ there are exactly m arcs going in (all of the form $xx_1x_2\cdots x_{n-1} \to x_1x_2\cdots x_n$, $x \in A_m$) and there are exactly m arcs coming out (all of the form $x_1x_2\cdots x_n \to x_2x_3\cdots x_nx$, $x \in A_m$). □

The following theorem shows that $\vec{D}_{n+1,m}$ is the line graph of $\vec{D}_{n,m}$. Recall the definition of the line graph in Definition 2.2.27.

Theorem 9.3.4. $\vec{D}_{n+1,m} = L(\vec{D}_{n,m})$.

Proof. Label each edge $x_1x_2\cdots x_n \to x_2x_3\cdots x_nx_{n+1}$ in $\vec{D}_{n,m}$ by $x_1x_2\cdots x_{n+1}$. Then it is not difficult to see that all arcs in $\vec{D}_{n,m}$ will be labelled in a bijective way by all the words of length $n+1$ over the alphabet A_m. Finally, a directed path of length 2

$$x_1x_2\cdots x_n \to x_2x_3\cdots x_nx_{n+1} \to x_3x_4\cdots x_nx_{n+1}x_{n+2}$$

in $\vec{D}_{n,m}$ produces an edge $x_1x_2\cdots x_{n+1} \to x_2x_3\cdots x_{n+1}x_{n+2}$, which is consistent with the definition of $\vec{D}_{n+1,m}$ and completes the proof that $\vec{D}_{n+1,m} = L(\vec{D}_{n,m})$. □

The existence of an Eulerian cycle in $\vec{D}_{n,m}$ shown in Theorem 9.3.3 gives the existence of a Hamiltonian cycle in $\vec{D}_{n+1,m}$ due to Theorem 9.3.4. Following an Eulerian cycle in $\vec{D}_{n,m}$, we can write down a word corresponding to it that will contain each word of length n over the alphabet A_m exactly once as a factor.

Example 9.3.5. The word 0001011100 is obtained from $\vec{D}_{3,2}$ following the Eulerian cycle

$$000 \to 001 \to 010 \to 101 \to 011 \to 111 \to 110 \to 100,$$

where we skipped recording the last step.

Thus, using graph theory we proved a result in combinatorics on words stating the existence of sequences containing each word of given length over a given alphabet as a factor exactly once. Such sequences are called *de Bruijn sequences* or, when the universal word is read cyclicly, *de Bruijn cycles*.

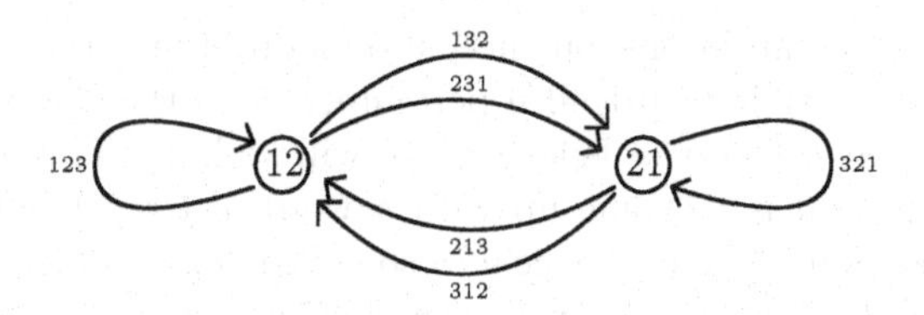

Figure 9.4: The graph $\mathcal{P}_2$ of overlapping permutations

9.3.2 Graphs of Overlapping Permutations

Graph of overlapping permutations are a permutation analogue of de Bruijn graphs.

Definition 9.3.6. A graph of overlapping permutations $\mathcal{P}_n$ has a directed edge for each permutation $\pi = \pi_1\pi_2\cdots\pi_{n+1} \in \mathcal{S}_{n+1}$, the set of all permutations of length $n+1$, from $\text{red}(\pi_1\pi_2\cdots\pi_n)$ to $\text{red}(\pi_2\pi_3\cdots\pi_{n+1})$.

Example 9.3.7. There are two arcs from 2341 to 3412 in $\mathcal{P}_4$ labelled 24513 and 34512. Indeed, $\text{red}(2451) = \text{red}(3451) = 2341$ and $\text{red}(4513) = \text{red}(4512) = 3412$. Also, see Figure 9.4 for the graph $\mathcal{P}_2$.

Graphs of overlapping permutations provide a useful tool for studying permutations. These graphs are yet another good example of interrelations between words (in this case permutations) and graphs.

9.3.3 Universal Cycles in Other Combinatorial Structures

Generalizing de Bruijn sequences/cycles considered in Subsection 9.3.1, we have the following definition.

Definition 9.3.8. Given a family C of combinatorial objects that can be represented by words of length n, a *universal cycle*, or *U-cycle*, for such a family is a word whose length-n factors, read cyclically, represent all the elements of C without repetition. Sometimes, the requirement for factors to be read cyclically is omitted, in which case we can talk about *universal words*, or *universal sequences*.

Example 9.3.9. Example 9.3.5 gives a universal word for the set of all binary words of length 3, which can be turned into a universal cycle for the same set by removing, for example, the last two 0s: 00010111.

Chung et al. [37] extended the study of de Bruijn sequences to other combinatorial structures, such as the set of all permutations, the set of all *partitions*, and the set of all *subsets of a finite set*.

In the case of permutations, one needs to modify the notion of a U-cycle since for $n \geq 3$ it is not possible to list all n-permutations in the way we treated words. Instead of looking at a factor of length n, we will look at the reduced form of the factor. Then, our goal is to come up with a word, a U-cycle, such that each n-permutation appears in it exactly once as a so-called *consecutive pattern*, that is, as the reduced form of a factor. The existence of such U-cycles for any n was proved in [37] by Chung et al. For example, for the set of length-3 permutations, such a U-cycle is 145243. Indeed, reading the reduced forms cyclically starting from the factor 145 we see the following permutations:

$$123 \to 231 \to 312 \to 132 \to 321 \to 213.$$

Another U-cycle for the same set is 142342. To list all length-n permutations, we need more than n different letters. Chung et al. [37] conjectured that $n+1$ letters are always enough for constructing a U-cycle for the set of length-n permutations for $n \geq 3$. This conjecture was proved in [83] by Johnson.

The existence of U-cycles for all permutations was proved in [37] by showing that the graph of overlapping permutations defined in Subsection 9.3.2 has a Hamiltonian cycle. Similarly, Burstein and Kitaev [30] proved the existence of U-cycles for all *word patterns* (which are the same as *multi-permutations*) by proving that a certain graph contains a Hamiltonian cycle.

See [81] by Jackson et al. for a recent collection of research problems (and references) on U-cycles.

9.4 Counting Independent Sets in Path-Schemes Using Combinatorics on Words

Let $G = (V, E)$ be a simple undirected graph with vertex set $V = \{1, \ldots, n\}$ and set of edges E.

The cardinality of $I(G)$, the set of all independent sets of G, was investigated in the literature for various classes of graphs, see, for example, [32, 62, 87]. In particular, Kitaev [87] used a classical result in combinatorics on words to find the generating function for the number of independent sets on the class of *well-based path-schemes* (see Theorem 9.4.14), which generalized a known result presented in Theorem 9.4.7. In this section, we discuss the results in [87] following closely the presentation in that paper. Moreover, in Subsection 9.4.5 we discuss well-based sequences associated with well-based path-schemes following [132] by Valyuzhenich.

9.4.1 Path-Schemes

Definition 9.4.1. Let M be a subset of $V = \{1, \ldots, n\}$. A *path-scheme* $P(n, M)$ is a graph $G = (V, E)$, where $E = \{xy \mid |x - y| \in M\}$.

See Figure 9.5 for an example of a path-scheme, where $n = 6$ and $M = \{2, 4\}$. In particular, in $P(6, \{2, 4\})$, there is an edge between the vertices 2 and 6 because $|2 - 6| = 4 \in M$ while the fact that $|2 - 5| = 3 \notin M$ tells us that there is no edge between the vertices 2 and 5.

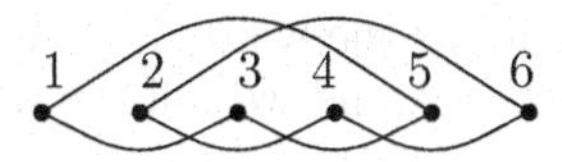

Figure 9.5: The path-scheme $P(6, \{2, 4\})$

By definition, $P(n, M)$ is a simple graph, and thus its adjacency matrix $A = (a_{i,j})_{n \times n}$ is symmetric. Moreover, if the columns and rows of A are ordered naturally, that is, vertex i corresponds to the ith column and the ith row, then for $1 \leq i < j < n$, $a_{i,j} = a_{i+1,j+1}$, since $|i - j|$ is in M if and only if $|(i + 1) - (j + 1)|$ is in M. Thus, we can construct the *upper triangular* part of A by shifting the first row to the right, that is, we place the first row, and row $i + 1$ is obtained by shifting row i one element to the right. Then we use symmetry to fill in the remaining entries of A. For example, for the path-scheme $P(6, \{2, 4\})$ in Figure 9.5, the adjacency matrix is

$$A = \begin{pmatrix} 0 & 0 & 1 & 0 & 1 & 0 \\ 0 & 0 & 0 & 1 & 0 & 1 \\ 1 & 0 & 0 & 0 & 1 & 0 \\ 0 & 1 & 0 & 0 & 0 & 1 \\ 1 & 0 & 1 & 0 & 0 & 0 \\ 0 & 1 & 0 & 1 & 0 & 0 \end{pmatrix}.$$

9.4.2 Well-Based Path-Schemes

Recall the definition of a factor in a word in Definition 3.2.8.

Definition 9.4.2. A word u *avoids a factor* v if u does not contain v as a factor.

Example 9.4.3. The word 24312 avoids the factor 23.

Definition 9.4.4. Suppose that $k \geq 2$ and $\mathcal{A} = \{A_1, \ldots, A_k\}$ is a set of words of the form $A_i = 1\underbrace{0\ldots0}_{a_i-1}1$, where $a_i \geq 1$, and $a_i < a_j$ if $i < j$. Moreover, we assume

Well-based set	Corresponding well-based sequence
$\{11, 101, 1001, \ldots, 10^{k-1}1\}$	$12\cdots k$
$\{11, 1001, 100001, \ldots, 10^{2m}1\}$	$135\cdots(2m+1)$
$\{11, 101, 1001, 1000001, 10000001\}$	12367

Table 9.1: Examples of well-based sets and well-based sequences

that for any $i > 1$ and $A_i \in \mathcal{A}$, if we replace any number of 0s in A_i by 1s, then we obtain a word A'_i that contains the word $A_j \in \mathcal{A}$ as a factor for some $j < i$. In this case, we call $\mathcal{A}$ a *well-based set*, and we call the sequence of a_is associated with $\mathcal{A}$ a *well-based sequence.*

Any well-based set must contain the word 11. Indeed, if we replace all 0s by 1s in, say, A_2 then A_1 must be a factor of the obtained word. So, we may extend our definition to the case $k = 1$. We define $\mathcal{A} = \{11\}$ to be a well-based set. We see that any well-based sequence starts from 1, and, clearly, if we take any number of consecutive initial elements of a well-based sequence, we obtain a well-based sequence.

Example 9.4.5. A few examples of well-based sets and the sequences associated with them are given in Table 9.1 (i copies of 0 is denoted by 0^i there).

Definition 9.4.6. We call a scheme $P(n, M)$ a *well-based scheme* if the elements of M listed in increasing order form a well-based sequence.

The following result is known and it is not difficult to prove.

Theorem 9.4.7. *(Folklore) One has that* $|I(P(n, \{1\}))| = F(n+2)$ *and, more generally,*

$$|I(P(n, \{1, \ldots, k-1\}))| = F_k(n+k), \tag{9.1}$$

where $F(n)$ *is the* n*th Fibonacci number and* $F_k(n)$ *is the* n*th* k-generalized Fibonacci number *defined as* $F_k(1) = \cdots = F_k(k) = 1$ *and*

$$F_k(n) = F_k(n-1) + F_k(n-k).$$

Note that in our notation, $F(n) = F_2(n)$.

In Subsection 9.4.4, we will provide the generating function for the number of independent sets of an arbitrary well-based path-scheme (see Theorem 9.4.14 obtained with the help of combinatorics on words). The results in Theorem 9.4.7 can be extracted from Theorem 9.4.14, since they correspond to a well-based sequence $1, \ldots, k-1$.

9.4.3 Counting Words Avoiding a Given Forbidden Set

Before going any further, we need some notions and results in a direction of combinatorics on words known as "binary factor avoidance". In our presentation we follow the master's thesis [134] of Bjorn Winterfjord, although the original ideas of respective derivations appear in [72].

A *binary word* is a word that consists of only the digits 0 and 1.

Definition 9.4.8. If $X_1 = a_0a_1 \cdots a_{k-1}$ and $X_2 = b_0b_1 \cdots b_{\ell-1}$ are two binary words of length k and ℓ respectively, then the *correlation* $c_{12} = c_0c_1 \cdots c_{k-1}$ between X_1 and X_2 viewed as an ordered pair (X_1, X_2) is the binary word defined as follows:

- if $k \leq \ell$: For all $0 \leq j \leq k-1$, $c_j = 1$ if $a_i = b_{\ell-k+i+j}$ for all $i = 0, 1, \ldots, k-j-1$, and $c_j = 0$ otherwise;
- if $k > \ell$: For all $0 \leq j \leq k - \ell$, $c_j = 1$ if $b_i = a_{k-\ell+i-j}$ for all $i = 0, 1, \ldots, \ell - 1$, and $c_j = 0$ otherwise; for all $k - \ell + 1 \leq j \leq k - 1$, $c_j = 1$ if $a_i = b_{\ell-k+i+j}$ for all $i = 0, 1, \ldots, k - j - 1$, and $c_j = 0$ otherwise.

Example 9.4.9. If $X_1 = 110$ and $X_2 = 1011$, then $c_{12} = 011$ and $c_{21} = 0010$, as depicted below.

$$\begin{array}{cccccccc} & & & 1 & 1 & 0 & & \\ \hline & & 1 & 0 & 1 & 1 & & 0 \\ & 1 & 0 & 1 & 1 & & & 1 \\ 1 & 0 & 1 & 1 & & & & 1 \end{array} \qquad \begin{array}{cccccccc} & & 1 & 0 & 1 & 1 & & \\ \hline & & & 1 & 1 & 0 & & 0 \\ & & 1 & 1 & 0 & & & 0 \\ & 1 & 1 & 0 & & & & 1 \\ 1 & 1 & 0 & & & & & 0 \end{array}$$

In general $c_{ij} \neq c_{ji}$ (they can even be of different lengths).

Definition 9.4.10. The *autocorrelation* of a word X_1 is just c_{11}, the correlation of X_1 with itself.

Example 9.4.11. If $X_1 = 1011$ then $c_{11} = 1001$. It is convenient to interpret a correlation $c_{ij} = c_0c_1 \cdots c_{k-1}$ as a polynomial $c_{ij}(x) = c_0 + c_1x + \cdots + c_{k-1}x^{k-1}$.

The following theorem is the main tool in proving Theorem 9.4.14. Recall from Definition 9.4.2 the notion of factor avoidance.

Theorem 9.4.12. *([134]) The generating function for the number of binary words that avoid the factors $b_1, b_2, \ldots, b_n$, of length $\ell_1, \ell_2, \ldots, \ell_n$ respectively, none included* *in any other, is given by the formula*

$$S_{b_1,b_2,\ldots,b_n}(x) = \frac{\begin{vmatrix} -c_{11}(x) & \cdots & -c_{1n}(x) \\ \vdots & \ddots & \vdots \\ -c_{n1}(x) & \cdots & -c_{nn}(x) \end{vmatrix}}{\begin{vmatrix} (1-2x) & 1 & \cdots & 1 \\ x^{\ell_1} & -c_{11}(x) & \cdots & -c_{1n}(x) \\ \vdots & \vdots & \ddots & \vdots \\ x^{\ell_n} & -c_{n1}(x) & \cdots & -c_{nn}(x) \end{vmatrix}}. \tag{9.2}$$

To demonstrate an application of Theorem 9.4.12 we present an example by Winterfjord.

Example 9.4.13. To find the generating function for the set of words that avoid the factors $b_1 = 111$, $b_2 = 110$ and $b_3 = 001$, we calculate that $c_{11}(x) = 1 + x + x^2$, $c_{12}(x) = 0$, $c_{13}(x) = x^2$, $c_{21}(x) = x + x^2$, $c_{22}(x) = 1$, $c_{23} = x^2$, $c_{31}(x) = 0$, $c_{32}(x) = x^2$ and $c_{33}(x) = 1$, and apply (9.2) to see that in this case

$$S_{111,110,001}(x) = \frac{1 + x + x^2 - x^4}{1 - x - x^2 + x^3} = 1 + 2x + 4x^2 + 5x^3 + 6x^4 + 7x^5 + 8x^6 + \cdots.$$

9.4.4 Counting Well-Based Path-Schemes

The following theorem is the main result in [87], and it is proved using a combinatorics on words approach.

Theorem 9.4.14. *([87]) Let $M = \{a_1, a_2, \ldots, a_k\}$ be a subset of $V = \{1, 2, \ldots, n\}$ such that the sequence $a_1, a_2, \ldots, a_k$ is well-based (in particular, $a_1 = 1$). Let $c(x) = 1 + \sum_{i=1}^{k} x^{a_i}$. Then the generating function for the number of independent sets in the well-based path-scheme $P = P(n, M)$ (with vertex set V) is given by*

$$G_M(x) = \frac{c(x)}{(1-x)c(x) - x}.$$

Even though we skip the proof of Theorem 9.4.14, we do discuss some corollaries to it as they appear in [87].

Example 9.4.15. If $M = \{1, 2, \ldots, k-1\}$ then we can apply Theorem 9.4.14, since the sequence $1, 2, \ldots, k-1$ is well-based. In this case, we obtain

$$G_{\{1,2,\ldots,k-1\}}(x) = \sum_{n \geq 0} g_n x^n = \frac{1 + x + \cdots + x^{k-1}}{1 - x - x^k},$$

and thus, using the form of the generating function, the sequence $g_n = |I(P(n, M))|$ satisfies the recurrence $g_n = g_{n-1} + g_{n-k}$ with $g_{1-k} = g_{2-k} = \cdots = g_0 = 1$, which agrees with (9.1).

Example 9.4.16. If $M = \{1, 3, 5\}$ then M is well-based. Theorem 9.4.14 gives us that

$$G_{\{1,3,5\}}(x) = \sum_{n \geq 0} w_n x^n = \frac{1 + x + x^3 + x^5}{1 - x - x^2 + x^3 - x^4 + x^5 - x^6}.$$

Thus, in this case the sequence w_n satisfies the recurrence formula

$$w_n = w_{n-1} + w_{n-2} - w_{n-3} + w_{n-4} - w_{n-5} + w_{n-6},$$

with the initial conditions: $w_{-5} = w_{-4} = w_{-3} = w_{-2} = w_{-1} = w_0 = 1$. The initial values for $w_n = |I(P(n, M))|$ and $n \geq 1$ are

$$2, 3, 5, 7, 11, 15, 23, 32, 49, 69, 105, 149, \ldots.$$

Finally, we state the following corollary to Theorem 9.4.14, which can be proved in a standard way by the partial fraction expansion of the generating function $G(x)$ from Theorem 9.4.14.

Corollary 9.4.17. *([87]) Let M, V, $c(x)$, and $P(n, M)$ satisfy the conditions of Theorem 9.4.14. Also, let ρ be the largest zero ($|\rho|$ is maximal among all the zeros) of the function*

$$Q(x) = (1 - x)c(x) - x = 1 - x - x^2 + (1 - x) \sum_{i=2}^{k} x^{a_i}.$$

Then, asymptotically, the growth rate of $|I(P(n, M))|$ is

$$|I(P(n, M))| \lesssim c|\rho|^n$$

for some constant c.

If $k = 1$ in Corollary 9.4.17 then $\rho = \frac{1+\sqrt{5}}{2}$, and if $k = 2$ then it can be shown that $0.6 \leq \rho \leq \frac{1+\sqrt{5}}{2}$.

9.4.5 On the Number of Well-Based Sequences

Theorem 9.4.14 gives a way to count independent sets in well-based path-schemes. However, to know how useful this theorem is, we want to know how large the class of well-based path-schemes is. Equivalently, how many different well-based sequences are there (see Definition 9.4.4)? While it does not seem to be feasible for the moment to find an exact enumeration of the sequences, it is possible to find its asymptotic growth. This subsection is based on the paper [132] by Valyuzhenich.

Clearly, in Definition 9.4.4 we can replace "any number of 0s in A_i by 1s" by "a 0 in A_i by 1" and still have an equivalent definition of a set. Thus, a well-based sequence $a_1a_2\cdots$ can alternatively be defined as follows: $a_1 = 1$ and for any $i \geq 2$, if $a_i = t + s$ for non-negative integers t and s, then either $t = a_j$ or $s = a_j$ for some $j < i$.

Let $W(n)$ be the set of well-based sequences whose elements are at most n, and let $w(n)$ be the number of sequences in $W(n)$.

Suppose that $a_1a_2\cdots$ is a sequence such that $a_i = i$ for $1 \leq i \leq k$ and $a_{k+1}a_{k+2}\cdots$ is obtained by listing an arbitrary subset of $\{k+2, k+3, \ldots, 2k+1\}$ in increasing order. It is not difficult to see that such a sequence is well-based (because all of the numbers in $\{1, \ldots, k\}$ are included in it) and there are 2^k such sequences. Moreover, taking different ks we obtain different sequences, since the lowest positive integers not included in these sequences, namely the $(k+1)$s, will be different. Thus, taking into account that the sequences $12\cdots(2n)$ and $12\cdots(2n+1)$ are well-based, which contributes the "1+" parts in the expressions below, we have

$$w(2n) \geq 1 + \sum_{i=0}^{n-1} 2^i = 2^n \quad \text{and} \quad w(2n+1) \geq 1 + \sum_{i=0}^{n} 2^i = 2^{n+1}.$$

Since the well-based sequence 135 was not included in our considerations above, we have that for $n \geq 5$,

$$w(n) > 2^{\lceil (n+1)/2 \rceil}. \tag{9.3}$$

The following notion was first discussed in [126, A103580].

Definition 9.4.18. A non-empty set $B \subseteq \{1, \ldots, n\}$ is a *basis* if no element x of B is a non-negative integer linear combination of the elements of $B - \{x\}$. Let $B(n)$ be the set of all such Bs, and $b(n)$ be the number of the elements in $B(n)$.

Example 9.4.19. All bases for $n = 4$ are $\{1\}$, $\{2\}$, $\{3\}$, $\{4\}$, $\{2,3\}$ and $\{3,4\}$.

Definition 9.4.20. A set S of integers is *sum-free* if for all $a, b \in S$, $a + b \notin S$. The family of all sum-free subsets $S \subseteq \{1, \ldots, n\}$ is denoted by $S(n)$, and $s(n)$ is the number of elements in $S(n)$.

Example 9.4.21. The set $\{3, 4, 8, 10\}$ is sum-free while the set $\{3, 5, 6, 8, 10\}$ is not sum-free because of the elements 3, 5 and 8.

It was shown in [132, Theorem 1] that there is a one-to-one correspondence between $W(n)$ and $B(n)$, so that $w(n) = b(n)$. Moreover, the following discussion was conducted in [132] (we refer to that paper for the respective references).

Note that any basis is a sum-free set. The converse to this statement does not hold, since, in particular, any subset of odd numbers is a sum-free set, but if its cardinality is at least 2, and it contains 1, then it is not a basis. Thus, $B(n) \subset S(n)$ implying that $w(n) = b(n) < s(n)$. Solving a conjecture of Cameron and Erdős, Green [71] and Sapozhenko [122] showed independently in 2003 that $s(n) = O(2^{n/2})$. Together with (9.3), we have the following result.

Theorem 9.4.22 ([132]). *The asymptotics for $w(n)$, the number of well-based sequences, is given by $\Theta(2^{n/2})$.*

9.5 Clique-Width of Graphs

The notion of clique-width of a graph was introduced in [43] and is defined as the minimum number of labels needed to construct the graph by means of the four graph operations:

- $i(v)$, the operation of creation of a vertex v with label i;
- $G \oplus H$, the disjoint union of two labelled graphs G and H;
- $\eta_{i,j}$, the operation that connects every vertex labelled i to every vertex labelled j;
- $\rho_{i \to j}$, the operation that renames label i to label j.

Every graph can be defined by an algebraic expression using the four operations above. For instance, the graph consisting of two adjacent vertices x and y can be defined by the expression $\eta_{1,2}(1(x) \oplus 2(y))$. Therefore, the notion of clique-width provides one more way of describing graphs by words. Moreover, any graph can be described by means of only the first three operations. In the worst case, such a description requires n labels for an n-vertex graph. However, for some graphs the use of the fourth operation allows the number of labels used in the description to be substantially reduced. For instance, it is not difficult to see that a path of any length can be described with the help of three different labels and a cycle of any length can be described with the help of four different labels. The following expression provides such a description for a cycle C_5 on vertices a, b, c, d, e (listed along the cycle):

$$\eta_{4,1}(\eta_{4,3}(4(e) \oplus \rho_{4\to 3}(\rho_{3\to 2}(\eta_{4,3}(4(d) \oplus \eta_{3,2}(3(c) \oplus \eta_{2,1}(2(b) \oplus 1(a)))))))).$$

An expression built from the above four operations is called a *k-expression* if it uses k different labels.

Definition 9.5.1. The *clique-width* of a graph G, denoted $cwd(G)$, is the minimum k such that there exists a k-expression defining G.

From the above examples we conclude that the clique-width of any path is at most 3 and the clique-width of any cycle is at most 4. The importance of the notion of clique-width is due to the fact that many algorithmic graph problems that are generally NP-hard admit polynomial-time solutions when restricted to graphs with bounded clique-width [42].

It is known that graphs of clique-width at most 2 are precisely P_4-free graphs (see, e.g. [44]). The set of graphs of clique-width at most 3 is substantially richer and includes, for instance, all forests. With k tending to infinity, graphs of clique-width at most k eventually cover all graphs, since the clique-width of any n-vertex graph is at most n. However, for any fixed k, the class of graphs of clique-width at most k is rather small and has at most factorial speed of growth defined as follows.

Definition 9.5.2. A class of graphs X is said to be *factorial* if the number X_n of n-vertex labelled graphs in X satisfies the inequalities $n^{c_1 n} \leq X_n \leq n^{c_2 n}$ for some constants c_1 and c_2.

In order to show that any class of graphs of bounded clique-width is at most factorial, let us observe that any algebraic expression defining a graph G can alternatively be represented as a rooted tree whose leaves correspond to the operations of vertex creation, whose internal nodes correspond to the $\oplus$-operations, and whose root is associated with G. The operations η and ρ are assigned to the respective edges of the tree. For example, the tree representing the expression above defining C_5 is depicted in Figure 9.6. Notice that any expression defining an n-vertex graph contains exactly $n-1$ $\oplus$-operations, and hence the corresponding tree has exactly $2n$ nodes.

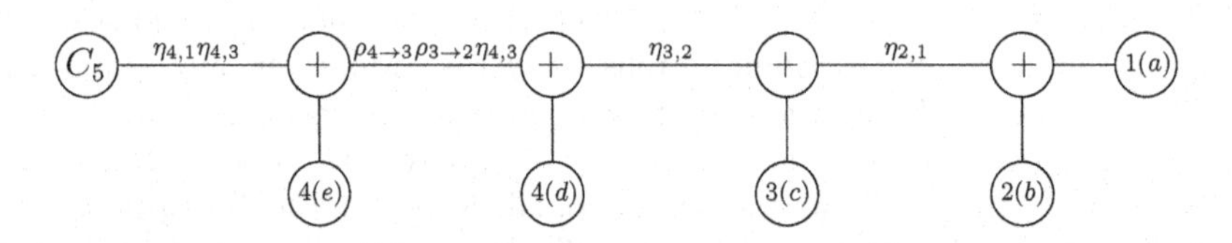

Figure 9.6: The tree representing the expression defining a C_5

We now use this representation together with the Prüfer sequence in order to prove the following theorem.

Theorem 9.5.3. *For any integer $k > 1$, the class of graphs with clique-width at most k, denoted by $\mathcal{C}(k)$, is factorial.*

Proof. $\mathcal{C}(2)$ contains the class of co-graphs, which is known to be factorial (see, e.g. [7]). This observation gives the lower bound. To obtain an upper bound on the number of graphs in $\mathcal{C}(k)$, we associate with each n-vertex graph $G \in \mathcal{C}(k)$ a k-expression $F(G)$ of minimum length and the tree $T(F(G))$ representing $F(G)$. The minimality of $F(G)$ ensures that the length of the label of any edge of the tree $T(F(G))$ is bounded by a constant depending on k but not n. Consequently, $T(F(G))$ can be encoded with a binary word of length $O(n \log_2 n)$, which immediately implies the required upper bound n^{cn} on the number of graphs in $\mathcal{C}(k)$.

An encoding of $T(F(G))$ of the desired length can be obtained, for instance, with the help of the Prüfer sequence. The tree $T(F(G))$ can easily be transformed into a labelled tree by assigning the labels of the edges to the respective parent nodes. Since $T(F(G))$ is a binary tree with $2n$ nodes and the length of any edge label is bounded by a constant, we obtain a labelled tree with node labels of length $O(\log_2 n)$ (in binary representation), as required. □

9.6 Conclusion

Quite often graphs are used as a tool for proving mathematical statements, in particular, statements in combinatorics on words. For example, finding the Gray code can be reduced to finding a Hamiltonian path in an n-cube, while finding de Bruijn sequences can be reduced to finding a Hamiltonian cycle in a de Bruijn graph. On the other hand, we do use combinatorics on words to prove statements in graph theory. The current chapter gives three such examples, namely, finding asymptotics for the snake-in-the-box problem, finding the number of independent sets of an arbitrary well-based path-scheme, and finding asymptotics for classes of graphs of bounded degree (via Prüfer sequences).

Appendix A

Graph Theory Background

A graph $G = (V, E)$ consists of two finite sets V and E. The elements of V are called the *vertices* and the elements of E the *edges* of G. Each edge is a pair of vertices.

Graphs have natural graphical representations in which each vertex is represented by a point and each edge by a line connecting two points. Figure A.1 represents the graph $G = (V, E)$ with vertex set $V = \{1, 2, 3, 4, 5\}$ and edge set $E = \{12, 23, 34, 45\}$.

1 2 3 4 5

Figure A.1: Graph $G = (V, E)$ with $V = \{1, 2, 3, 4, 5\}$ and $E = \{12, 23, 34, 45\}$

By altering the definition, we can obtain different types of graphs. For instance,

- by replacing the set E with a set of *ordered pairs* of vertices, we obtain a *directed graph* (*digraph* for short), also known as an *oriented graph*.
- by allowing E to contain both directed and undirected edges, we obtain a *mixed graph*.
- by allowing repeated elements in the set of edges, i.e., by replacing E with a multiset, we obtain a *multigraph*.
- by allowing edges to connect a vertex to itself (a loop), we obtain a *pseudograph*.
- by allowing the edges to be arbitrary subsets of vertices, not necessarily of size two, we obtain a *hypergraph*.

- by allowing V and E to be infinite sets, we obtain an *infinite graph.*

Definition A.0.1. A *simple graph* is a finite undirected graph without loops and multiple edges.

All graphs in this book are simple, unless stated otherwise.

A.1 Terminology and Notation

For a graph G, we denote by $V(G)$ and $E(G)$ the set of vertices and the set of edges of G, respectively.

Definition A.1.1. Let u, v be two vertices of a graph G.

- If $uv \in E(G)$, then u, v are said to be *adjacent*, in which case we also say that u is *connected* to v or u is a *neighbour* of v. If $uv \notin E(G)$, then u and v are *nonadjacent* (*not connected, non-neighbours*).
- The *neighbourhood* of a vertex $v \in V(G)$, denoted $N(v)$, is the set of vertices adjacent to v, i.e. $N(v) = \{u \in V(G) \mid vu \in E(G)\}$.
- If $e = uv$ is an edge of G, then e is *incident* to u and v. We also say that u and v are the *endpoints* of e.
- The *degree* of $v \in V(G)$, denoted $deg(v)$, is the number of edges incident to v. Alternatively, $deg(v) = |N(v)|$.

Definition A.1.2. The *complement* of a graph $G = (V, E)$ is a graph with vertex set V and edge set E' such that $e \in E'$ if and only if $e \notin E$. The complement of a graph G is denoted $\overline{G}$ or co-G.

Definition A.1.3. Given two graphs $G_1 = (V_1, E_1)$ and $G_2 = (V_2, E_2)$, the graph G_1 is said to be

- a *subgraph* of G_2 if $V_1 \subseteq V_2$ and $E_1 \subseteq E_2$, i.e. G_1 can be obtained from G_2 by deleting some vertices (with all incident edges) and some edges;
- a *spanning subgraph* of G_2 if $V_1 = V_2$, i.e. G_1 can be obtained from G_2 by deleting some edges but not vertices;
- an *induced subgraph* of G_2 if every edge of G_2 with both endpoints in V_1 is also an edge of G_1, i.e. G_1 can be obtained from G_2 by deleting some vertices (with all incident edges) but not edges.

Definition A.1.4. In a graph,

- a *walk* is a sequence of (not necessarily distinct) vertices $v_1, v_2, \ldots, v_k$ such that v_i is adjacent to v_{i+1} for each $i = 1, \ldots, k-1$. A walk can alternatively be described by the set of its edges $v_1v_2, \ldots, v_{k-1}v_k$. The *length* of a walk is the number of its edges. A walk is *closed* if $v_1 = v_k$.
- a *trail* is a walk with no repeated edges.
- a *path* is a walk with no repeated vertices.
- a *Hamiltonian path* is a path containing all vertices of the graph.
- a *cycle* is a closed walk with no repeated vertices (except for $v_1 = v_k$).
- a *Hamiltonian cycle* is a cycle containing all vertices of the graph.
- an *Eulerian cycle* is a walk containing every edge of the graph exactly once.
- the *distance* between two vertices a and b, denoted $dist(a, b)$, is the length of a shortest path joining them.

Definition A.1.5. A graph G is called *connected* if every pair of distinct vertices is joined by a path. Otherwise it is disconnected. A maximal (with respect to inclusion) connected subgraph of G is called a *connected component* of G.

Definition A.1.6. The *diameter* of a graph G is the maximum distance between two of G's vertices. A disconnected graph has infinite diameter.

Definition A.1.7. Two graphs $G_1 = (V_1, E_1)$ and $G_2 = (V_2, E_2)$ are *isomorphic* if there is a bijection $f : V_1 \to V_2$ that preserves adjacency, i.e., $uv \in E_1$ if and only if $f(u)f(v) \in E_2$.

Definition A.1.8. A graph G is *self-complementary* if G is isomorphic to its complement.

Definition A.1.9. In a graph,

- a *clique* is a set of pairwise adjacent vertices. The size of a maximum clique in G is called the *clique number* of G and is denoted $\omega(G)$.
- an *independent set* (also known as a *stable set*) is a set of pairwise non-adjacent vertices. The size of a maximum independent set in G is called the *independence number* (also known as the *stability number*) of G and is denoted $\alpha(G)$.

- a *matching* is a set of edges no two of which share a vertex. A matching is *perfect* if it covers all vertices of the graph.

Definition A.1.10. A *bipartite graph* is a graph whose vertices can be divided into two disjoint sets U and V such that every edge connects a vertex in U to one in V.

Some commonly used graphs:

- K_n is the *complete graph* with n vertices, i.e. the graph with n vertices every two of which are adjacent.
- $K_{n,m}$ is the *complete bipartite graph*, i.e. a bipartite graph with parts U and V of size n and m, respectively, such that each vertex in U is connected to every vertex in V.
- O_n is the *edgeless graph* on n vertices, i.e. the graph on n vertices that has no edges.
- P_n is a chordless *path* with n vertices, i.e. $V(P_n) = \{v_1, v_2, \ldots, v_n\}$ and $E(P_n) = \{v_1v_2, \ldots, v_{n-1}v_n\}$.
- C_n is a chordless *cycle* with n vertices, i.e. $V(C_n) = \{v_1, v_2, \ldots, v_n\}$ and $E(C_n) = \{v_1v_2, \ldots, v_{n-1}v_n, v_nv_1\}$.
- W_n is a *wheel*, i.e. a graph obtained from a cycle C_n by adding a vertex which is adjacent to every vertex of the cycle.
- $G + H$ is the union of two disjoint graphs G and H, i.e. if $V(G) \cap V(H) = \emptyset$, then $V(G+H) = V(G) \cup V(H)$ and $E(G+H) = E(G) \cup E(H)$. In particular, nG denotes the disjoint union of n copies of G.

Definition A.1.11. A *star* or *star tree* S_n is the complete bipartite graph $K_{1,n}$ where $n \geq 0$ and $n = 0$ corresponds to the single-vertex graph K_1. The *centrum* of a star is the all-adjacent vertex in it.

Definition A.1.12. The *Ramsey number* $R(n, m)$ gives the solution to the party problem, which asks the minimum number of guests that must be invited so that at least m will know each other or at least n will not know each other. In the language of graph theory, the Ramsey number is the minimum number of vertices $v = R(n, m)$ such that all undirected simple graphs of order v contain a clique of order n or an independent set of order m.

The well known *Ramsey's Theorem* states that such a number exists for all n and m.

A.2 Directed Graphs

In this section, we present some terminology and notation for directed graphs. We repeat that in a directed graph every edge is an *ordered* pair of vertices. Frequently, directed edges are called *arcs*. Below, we represent an arc from a to b by (a, b), although in the book, we also use the notation $a \to b$. Graphically, the orientation of an arc is indicated by an arrow (see Figure A.2). Observe that two vertices of an oriented graph can be connected by *two* arcs directed opposite to each other.

1 2 3 4 5

Figure A.2: An oriented graph $G = (V, E)$ with $V = \{1, 2, 3, 4, 5\}$ and $E = \{(1, 2), (3, 2), (3, 4), (5, 4)\}$

Definition A.2.1. Let G be a directed graph and v a vertex of G.

- The *in-degree* of v is the number of edges directed *to* v, i.e. of the form (u, v).
- The *out-degree* of v is the number of edges directed *out* of v, i.e. of the form (v, u).
- G is *balanced* if the in-degree equals the out-degree for each vertex of the graph.

Definition A.2.2. A directed graph is *strongly connected* if every vertex of the graph is connected to every other vertex of the graph by a *directed path* (i.e. a path that respects the orientation of the edges). A directed graph is *weakly connected* if the underlying graph is connected.

A.3 Some Graph Theory Results Used in the Book

In this section we provide two well known results in graph theory.

Theorem A.3.1. *A graph contains an Eulerian cycle if and only if the graph is connected and the degree of each of its vertices is even.*

Theorem A.3.2. *A digraph contains an Eulerian cycle if and only if it is balanced and strongly connected.*

Appendix B

Non-graph-theoretical Background

Here we provide definitions of some basic notions used in the book.

B.1 Algebra and Analysis

Definition B.1.1. For sets A and B, the *Cartesian product* $A \times B$ is the set of all ordered pairs (a, b) where $a \in A$ and $b \in B$. That is,

$$A \times B = \{(a, b) \mid a \in A \text{ and } b \in B\}.$$

Definition B.1.2. $\mathcal{S}_n$ denotes the set of all permutations of length n. A permutation of length n is normally thought of by us as a word of length n, without repeated letters, over an n-letter alphabet, typically $\{1, 2, \ldots, n\}$.

Definition B.1.3. A (non-strict) *partial order* is a binary relation "$\leq$" over a set P which is *reflexive*, *antisymmetric* and *transitive*, i.e. which satisfies for all a, b and c in P the following conditions:

- $a \leq a$ (reflexivity);
- if $a \leq b$ and $b \leq a$ then $a = b$ (antisymmetry);
- if $a \leq b$ and $b \leq c$ then $a \leq c$ (transitivity).

Definition B.1.4. A set with a partial order is called a *partially ordered set*, or a *poset*.

Definition B.1.5. For a, b, elements of a partially ordered set P, if $a \leq b$ or $b \leq a$, then a and b are *comparable*. Otherwise they are *incomparable*.

Definition B.1.6. A partial order under which every pair of elements is comparable is called a *total order* or *linear order.* A totally ordered set is also called a *chain.*

Definition B.1.7. An element a is said to be *covered* by another element b if a is strictly less than b and no third element c fits between them. That is, both $a \leq b$ and $a \neq b$ are true, and $a \leq c \leq b$ is false for each c with $a \neq c \neq b$.

Definition B.1.8. A *linear extension* of a partial order is a linear order that is compatible with the partial order.

Definition B.1.9. The *dimension* of a poset P is the least integer t for which there exists a family of t linear extensions of P so that, for every x and y in P, x precedes y in P if and only if it precedes y in each of the linear extensions.

Definition B.1.10. A family $\mathcal{R} = (\leq_1, \ldots, \leq_t)$ of linear orders on a set X is called a *realizer* of a poset $P = (X, \leq_P)$ if $\leq_p = \cap\mathcal{R}$ which is to say that for any x and y in X, $x \leq_P y$ precisely when $x \leq_1 y$, $x \leq_2 y, \ldots$, and $x \leq_t y$.

Definition B.1.11. Suppose that S is a set and $\cdot$ is some binary operation $S \times S \to S$, then S with $\cdot$ is a *monoid* if it satisfies the following two axioms:

- **Associativity.** For all a, b and c in S, the equation $(a \cdot b) \cdot c = a \cdot (b \cdot c)$ holds.
- **Identity element.** There exists an elenent e in S such that for every element a in S, the equations $e \cdot a = a \cdot e = a$ hold.

Definition B.1.12. For a set A, the set of all words over A, denoted A^*, is called the *free monoid.*

Definition B.1.13. Given a set A, the *free commutative monoid* on A, denoted $[A]$, is the set of all finite multisets with elements drawn from A, with the monoid operation being multiset sum and the monoid unit being the empty multiset.

A *morphism* is a structure-preserving map from one structure to another. A special case of a morphism is defined next.

Definition B.1.14. A map $f : X \to Y$ is called an *epimorphism* if $g_1 \cdot f = g_2 \cdot f$ implies $g_1 = g_2$ for all morphisms $g_1, g_2 : Y \to Z$.

In the following definition, one may think of a *formal power series* as a power series in which we ignore questions of convergence by not assuming that the variable denotes any numerical value.

Definition B.1.15. A *generating function* is a formal power series in one indeterminate, whose coefficients encode information about a sequence of numbers a_n that is indexed by the natural numbers. The *ordinary generating function*, or just the generating function, of a sequence a_n is $\sum_{n=0}^{\infty} a_n x^n$.

Generating functions are a very powerful tool in modern mathematics, in particular in combinatorics.

We end this section by presenting three members of the *family of Bachmann-Landau notations* used in this book.

Definition B.1.16 (Big-O notation). $f(x) = O(g(x))$ means that there exists a positive real number M and a real number x_0 such that $|f(x)| \leq M \cdot |g(x)|$ for all $x \geq x_0$.

Definition B.1.17 (Small-O notation). $f(x) = o(g(x))$ means that for all $c > 0$ there exists some $k > 0$ such that $0 \leq f(n) < c \cdot g(n)$ for all $n \geq k$. The value of k must not depend on n, but may depend on c.

Definition B.1.18 (Big-Theta notation). $f(n) = \Theta(g(n))$ if f is bounded both above and below by g asymptotically. That is, there exist positive k_1, k_2 and n_0 such that $k_1 \cdot g(n) \leq f(n) \leq k_2 \cdot g(n)$ for all $n > n_0$.

B.2 Combinatorics

Definition B.2.1. The sequence $F(n)$ of *Fibonacci numbers* is defined by the recurrence relation $F(n) = F(n-1) + F(n-2)$ with $F(0) = 0$ and $F(1) = 1$.

Fibonacci numbers, which have numerous applications, are probably the best known, and definitely the oldest known (dating back to ca. 200BC) sequence of numbers. The first few Fibonacci numbers, for $n = 0, 1, 2, \ldots$, are

$$0, 1, 1, 2, 3, 5, 8, 13, 21, 34, 55, 89, 144, \ldots.$$

Definition B.2.2. The *nth Catalan number* is $\frac{1}{n+1}\binom{2n}{n}$.

The Catalan numbers occur in many counting problems, often involving recursively defined structures. The first few Catalan numbers, for $n = 0, 1, 2, \ldots$, are

$$1, 1, 2, 5, 14, 42, 132, 429, 1430, 4862, 16796, 58786, 208012, 742900, \ldots.$$

Definition B.2.3. A *Stirling number of the second kind* is the number of ways to partition a set of n objects into k non-empty subsets and is denoted by $S(n, k)$ or $\left\{ n \atop k \right\}$.

The initial values for the Stirling numbers of the second kind are presented in Table B.1.

$n\backslash k$	0	1	2	3	4	5	6	7
0	1							
1	0	1						
2	0	1	1					
3	0	1	3	1				
4	0	1	7	6	1			
5	0	1	15	25	10	1		
6	0	1	31	90	65	15	1	
7	0	1	63	301	350	140	21	1

Table B.1: The initial values for the Stirling numbers of the second kind

Definition B.2.4. The nth *Bell number* B_n counts the number of different ways to partition a set with n elements.

The first few Bell numbers, for $n = 0, 1, 2, \ldots$, are

$$1, 1, 2, 5, 15, 52, 203, 877, 4140, 21147, 115975, \ldots.$$

The following proposition is known as the *pigeonhole principle.* Despite being trivial, the principle has found many applications in various problems, e.g. in proving that the maximum number of non-attacking knights on the chessboard is 32.

Proposition B.2.5 (The pigeonhole principle). *If n items are put into m containers, with $n > m$, then at least one container must contain more than one item.*

Bibliography

[1] Jaromír Abrham and Anton Kotzig. Transformations of Euler tours. *Ann. Discrete Math.*, 8:65–69, 1980. Combinatorics 79 (Proc. Colloq., Univ. Montréal, Montreal, 1979), Part I.

[2] Martin Aigner and Günter M. Ziegler. *Proofs from The Book*. Springer, fourth edition, 2010.

[3] Prosper Akrobotu, Sergey Kitaev, and Zuzana Masárová. On word-representability of polyomino triangulations. *Siberian Adv. in Math.*, 25(1):1–10, 2015.

[4] M.H. Albert, M. D. Atkinson, M. Bouvel, N. Ruškuc, and V. Vatter. Geometric grid classes of permutations. *Trans. Amer. Math. Soc.*, 365:5859–5881, 2013.

[5] V. E. Alekseev. Hereditary classes and coding of graphs. *Problemy Kibernet.*, (39):151–164, 1982.

[6] V. E. Alekseev. Range of values of entropy of hereditary classes of graphs. *Diskret. Mat.*, 4(2):148–157, 1992.

[7] V. E. Alekseev. On lower layers of a lattice of hereditary classes of graphs. *Diskretn. Anal. Issled. Oper. Ser. 1*, 4(1):3–12, 1997. In Russian.

[8] Peter Allen, Vadim Lozin, and Michaël Rao. Clique-width and the speed of hereditary properties. *Electron. J. Combin.*, 16(1):Research Paper 35, 11 pp, 2009.

[9] Noga Alon, József Balogh, Béla Bollobás, and Robert Morris. The structure of almost all graphs in a hereditary property. *J. Combin. Theory Ser. B*, 101(2):85–110, 2011.

[10] B. Alspach and N. J. Pullman. Path decomposition of digraphs. *Bulletin of the Australian Mathematical Society*, 10:421–427, 1974.

[11] Stephen Alstrup and Theis Rauhe. Small induced-universal graphs and compact implicit graph representations. In *Foundations of Computer Science, 2002. Proceedings. The 43rd Annual IEEE Symposium on*, pages 53–62, 2002.

[12] M. D. Atkinson, M. M. Murphy, and N. Ruškuc. Partially well-ordered closed sets of permutations. *Order*, 19(2):101–113, 2002.

[13] Aistis Atminas, Andrew Collins, Jan Foniok, and Vadim V. Lozin. Deciding the Bell number for hereditary graph properties. In Dieter Kratsch and Ioan Todinca, editors, *Graph-Theoretic Concepts in Computer Science*, volume 8747 of *Lecture Notes in Computer Science*, pages 69–80. Springer, 2014.

[14] Aistis Atminas, Vadim V. Lozin, Sergey Kitaev, and Alexandr Valyuzhenich. Universal graphs and universal permutations. *Discrete Math. Algorithms Appl.*, 5(4):1350038, 15, 2013.

[15] K. A. Baker, P. Fishburn, and F. S. Roberts. Partial orders of dimension 2. *Networks*, 2(1):1128, 1971.

[16] József Balogh, Béla Bollobás, and David Weinreich. The speed of hereditary properties of graphs. *J. Combin. Theory Ser. B*, 79(2):131–156, 2000.

[17] József Balogh, Béla Bollobás, and David Weinreich. The penultimate rate of growth for graph properties. *European J. Combin.*, 22(3):277–289, 2001.

[18] József Balogh, Béla Bollobás, and David Weinreich. A jump to the Bell number for hereditary graph properties. *J. Combin. Theory Ser. B*, 95(1):29–48, 2005.

[19] Michael J. Bannister, Zhanpeng Cheng, William E. Devanny, and David Eppstein. Superpatterns and universal point sets. *J. Graph Algorithms Appl.*, 18(2):177–209, 2014.

[20] Michael J. Bannister, William E. Devanny, and David Eppstein. Small superpatterns for dominance drawing. In Michael Drmota and Mark Daniel Ward, editors, *2014 Proceedings of the Eleventh Workshop on Analytic Algorithmics and Combinatorics (ANALCO)*, pages 92–103. SIAM, Philadelphia, 2014.

[21] Lowell W. Beineke. Characterizations of derived graphs. *Journal of Combinatorial Theory*, 9(2):129 – 135, 1970.

[22] Edward J. L. Bell. Word-graph theory. *Lancaster University*, 2011.

[23] Edward J. L. Bell, Paul Rayson, and Damon Berridge. The strong-connectivity of word-representable digraphs. *arXiv:1102.0980 [math.CO]*, 2011.

[24] Béla Bollobás and Andrew Thomason. Projections of bodies and hereditary properties of hypergraphs. *Bull. London Math. Soc.*, 27(5):417–424, 1995.

[25] Béla Bollobás and Andrew Thomason. Hereditary and monotone properties of graphs. In *The Mathematics of Paul Erdős, II*, volume 14 of *Algorithms Combin.*, pages 70–78. Springer, 1997.

[26] André Bouchet. Circle graph obstructions. *J. Combin. Theory Ser. B*, 60(1):107–144, 1994.

[27] Andreas Brandstädt, Van Bang Le, and Jeremy P. Spinrad. *Graph classes: a survey.* SIAM Monographs on Discrete Mathematics and Applications. Society for Industrial and Applied Mathematics (SIAM), Philadelphia, 1999.

[28] Graham Brightwell. On the complexity of diagram testing. *Order*, 10(4):297–303, 1993.

[29] R. L. Brooks. On colouring the nodes of a network. *Proc. Cambridge Philos. Soc.*, 37:194–197, 1941.

[30] A. Burstein and S. Kitaev. On unavoidable sets of word patterns. *SIAM J. Discrete Math.*, 19(2):371–381 (electronic), 2005.

[31] Steve Butler. Induced-universal graphs for graphs with bounded maximum degree. *Graphs Combin.*, 25(4):461–468, 2009.

[32] Neil J. Calkin and Herbert S. Wilf. The number of independent sets in a grid graph. *SIAM J. Discret. Math.*, 11(1):54–60, February 1998.

[33] Parinya Chalermsook, Bundit Laekhanukit, and Danupon Nanongkai. Graph products revisited: Tight approximation hardness of induced matching, poset dimension and more. In *Proceedings of the Twenty-Fourth Annual ACM-SIAM Symposium on Discrete Algorithms*, SODA '13, pages 1557–1576. SIAM, Philadelphia, 2013.

[34] Thomas Z. Q. Chen. Private communication. 2015.

[35] Gregory L. Cherlin and Brenda J. Latka. Minimal antichains in well-founded quasi-orders with an application to tournaments. *J. Combin. Theory Ser. B*, 80(2):258–276, 2000.

[36] Maria Chudnovsky, Neil Robertson, Paul Seymour, and Robin Thomas. The strong perfect graph theorem. *Ann. of Math. (2)*, 164(1):51–229, 2006.

[37] F. R. K. Chung, P. Diaconis, and R. Graham. Universal cycles for combinatorial structures. *Discrete Math.*, 110(1–3):43–59, 1992.

[38] Fan R. K. Chung. Universal graphs and induced-universal graphs. *J. Graph Theory*, 14(4):443–454, 1990.

[39] Václav Chvátal and Peter L. Hammer. Aggregation of inequalities in integer programming. In *Studies in Integer Programming (Proc. Workshop, Bonn, 1975)*, pages 145–162. Ann. of Discrete Math., Vol. 1. North-Holland, Amsterdam, 1977.

[40] Andrew Collins, Sergey Kitaev, and Vadim Lozin. New results on word-representable graphs. *Discrete Applied Mathematics, to appear*, 2015.

[41] D. G. Corneil, H. Lerchs, and L. Stewart Burlingham. Complement reducible graphs. *Discrete Appl. Math.*, 3(3):163–174, 1981.

[42] B. Courcelle, J. A. Makowsky, and U. Rotics. Linear time solvable optimization problems on graphs of bounded clique-width. *Theory Comput. Syst.*, 33(2):125–150, 2000.

[43] Bruno Courcelle, Joost Engelfriet, and Grzegorz Rozenberg. Handle-rewriting hypergraph grammars. *J. Comput. System Sci.*, 46(2):218–270, 1993.

[44] Bruno Courcelle and Stephan Olariu. Upper bounds to the clique width of graphs. *Discrete Appl. Math.*, 101(1-3):77–114, 2000.

[45] Peter Damaschke. Induced subgraphs and well-quasi-ordering. *J. Graph Theory*, 14(4):427–435, 1990.

[46] Hubert de Fraysseix. Local complementation and interlacement graphs. *Discrete Math.*, 33(1):29–35, 1981.

[47] Aldo de Luca and Stefano Varricchio. Well quasi-orders and regular languages. *Acta Inform.*, 31(6):539–557, 1994.

[48] R. W. Doran. The Gray code. *J.UCS*, 13(11):1573–1597, 2007.

[49] Ben Dushnik and E. W. Miller. Partially ordered sets. *Amer. J. Math.*, 63:600–610, 1941.

[50] Paul Erdős. Graph theory and probability. *Canad. J. Math.*, 11:34–38, 1959.

[51] Paul Erdős, Peter Frankl, and Vojtěch Rödl. The asymptotic number of graphs not containing a fixed subgraph and a problem for hypergraphs having no exponent. *Graphs Combin.*, 2(2):113–121, 1986.

[52] Paul Erdős, Daniel J. Kleitman, and Bruce L. Rothschild. Asymptotic enumeration of K_n-free graphs. In *Colloquio Internazionale sulle Teorie Combinatorie (Rome, 1973), Volo II*, number 17 in Atti dei Convegni Lincei, pages 19–27. Accad. Naz. Lincei, Rome, 1976.

[53] Paul Erdős and George Szekeres. A combinatorial problem in geometry. *Compositio Math.*, 2:463–470, 1935.

[54] Henrik Eriksson, Kimmo Eriksson, Svante Linusson, and Johan Wästlund. Dense packing of patterns in a permutation. *Ann. Comb.*, 11(3-4):459–470, 2007.

[55] Louis Esperet, Arnaud Labourel, and Pascal Ochem. On induced-universal graphs for the class of bounded-degree graphs. *Inform. Process. Lett.*, 108(5):255–260, 2008.

[56] A. A. Evdokimov. The maximal length of a chain in the unit n-dimensional cube. *Mat. Zametki*, 6:309–319, 1969.

[57] Michael R. Fellows. The Robertson-Seymour theorems: a survey of applications. In *Graphs and Algorithms (Boulder, 1987)*, volume 89 of *Contemp. Math.*, pages 1–18. Amer. Math. Soc., Providence, 1989.

[58] Cristina G. Fernandes, Edward L. Green, and Arnaldo Mandel. From monomials to words to graphs. *J. Combin. Theory Ser. A*, 105(2):185–206, 2004.

[59] A. Finkel and Ph. Schnoebelen. Well-structured transition systems everywhere! *Theoret. Comput. Sci.*, 256(1-2):63–92, 2001.

[60] Peter C. Fishburn. An interval graph is not a comparability graph. *J. Combinatorial Theory*, 8(4):442–443, 1970.

[61] Stéphane Foldes and Peter L. Hammer. Split graphs. In *Proceedings of the Eighth Southeastern Conference on Combinatorics, Graph Theory and Computing (Louisiana State Univ., Baton Rouge, 1977)*, pages 311–315. Congressus Numerantium, No. XIX.

[62] Florence Forbes and Bernard Ycart. Counting stable sets on Cartesian products of graphs. *Discrete Mathematics*, 186(1–3):105–116, 1998.

[63] T. Gallai. Transitiv orientierbare Graphen. *Acta Math. Acad. Sci. Hungar*, 18:25–66, 1967.

[64] Hana Galperin and Avi Wigderson. Succinct representations of graphs. *Inform. and Control*, 56(3):183–198, 1983.

[65] Ian P. Gent, Christopher Jefferson, and Ian Miguel. Minion: A fast scalable constraint solver. In Gerhard Brewka, Silvia Coradeschi, Anna Perini, and Paolo Traverso, editors, *ECAI 2006, 17th European Conference on Artificial Intelligence, August 29 - September 1, 2006, Riva del Garda, Italy, including Prestigious Applications of Intelligent Systems (PAIS 2006), Proceedings*, volume 141 of *Frontiers in Artificial Intelligence and Applications*, pages 98–102. IOS Press, Amsterdam, 2006.

[66] P. C. Gilmore and A. J. Hoffman. A characterization of comparability graphs and of interval graphs. *Canad. J. Math.*, 16:539–548, 1964.

[67] Marc Glen. Word-representable graphs. *Software available at* `personal.cis.strath.ac.uk/sergey.kitaev/word-representable-graphs.html`, 2015.

[68] Marc Glen and Sergey Kitaev. Word-representability of triangulations of rectangular polyomino with a single domino tile. *Journal of Combinatorial Mathematics and Combinatorial Computing, to appear*, 2015.

[69] Martin Charles Golumbic. *Algorithmic Graph Theory and Perfect Graphs*, volume 57 of *Annals of Discrete Mathematics*. Elsevier Science B.V., Amsterdam, second edition, 2004. With a foreword by Claude Berge.

[70] Ron Graham and Nan Zang. Enumerating split-pair arrangements. *J. Combin. Theory Ser. A*, 115(2):293–303, 2008.

[71] Ben Green. The cameron-Erdös conjecture. *Bull. London Math. Soc.*, 36(6):769–778, 2004.

[72] Leo J. Guibas and Andrew M. Odlyzko. Periods in strings. *J. Combin. Theory Ser. A*, 30(1):19–42, 1981.

[73] Magnús M. Halldórsson, Sergey Kitaev, and Artem Pyatkin. Graphs capturing alternations in words. In Yuan Gao, Hanlin Lu, Shinnosuke Seki, and Sheng Yu, editors, *Developments in Language Theory*, volume 6224 of *Lecture Notes in Computer Science*, pages 436–437. Springer Berlin, 2010.

[74] Magnús M. Halldórsson, Sergey Kitaev, and Artem Pyatkin. Alternation graphs. In Petr Kolman and Jan Kratochvíl, editors, *Graph-Theoretic Concepts in Computer Science*, volume 6986 of *Lecture Notes in Computer Science*, pages 191–202. Springer Berlin Heidelberg, 2011.

[75] Magnús M. Halldórsson, Sergey Kitaev, and Artem Pyatkin. Semi-transitive orientations and word-representable graphs. *Discrete Applied Mathematics, to appear*, 2015.

[76] P. L. Hammer and A. K. Kelmans. On universal threshold graphs. *Combin. Probab. Comput.*, 3(3):327–344, 1994.

[77] Frank Harary. *Graph theory.* Addison-Wesley, Reading, 1969.

[78] Rajneesh Hegde and Kamal Jain. The hardness of approximating poset dimension. *Electronic Notes in Discrete Mathematics*, 29(0):435 – 443, 2007. European Conference on Combinatorics, Graph Theory and Applications.

[79] Graham Higman. Ordering by divisibility in abstract algebras. *Proc. London Math. Soc. (3)*, 2(1):326–336, 1952.

[80] Natalie Hine and James Oxley. When excluding one matroid prevents infinite antichains. *Adv. in Appl. Math.*, 45(1):74–76, 2010.

[81] B. Jackson, B. Stevens, and G. Hurlbert. Research problems on Gray codes and universal cycles. *Discrete Math.*, 309(17):5341–5348, 2009.

[82] Michel Jean. An interval graph is a comparability graph. *J. Combinatorial Theory*, 7(2):189–190, 1969.

[83] J. R. Johnson. Universal cycles for permutations. *Discrete Math.*, 309(17):5264–5270, 2009.

[84] Miles Jones, Sergey Kitaev, Artem Pyatkin, and Jeffrey Remmel. Representing graphs via pattern avoiding words. *Electron. J. Combin.*, 22(2):Paper 2.53, 20, 2015.

[85] Sampath Kannan, Moni Naor, and Steven Rudich. Implicit representation of graphs. *SIAM J. Discrete Math.*, 5(4):596–603, 1992.

[86] W. H. Kautz. Unit-distance error-checking codes. *IRE Transactions on Electronic Computers*, 7(2):177–180, 1958.

[87] Sergey Kitaev. Independent sets on path-schemes. *J. Integer Seq.*, 9(2):Article 06.2.2, 8, 2006.

[88] Sergey Kitaev. *Patterns in Permutations and Words.* Springer, 2011.

[89] Sergey Kitaev. On graphs with representation number 3. *arXiv:1403.1616*, 2014.

[90] Sergey Kitaev. Existence of u-representation of graphs. *arXiv: 1507.03177*, 2015.

[91] Sergey Kitaev, Toufik Mansour, and Patrice Séébold. Generating the Peano curve and counting occurrences of some patterns. *J. Autom. Lang. Comb.*, 9(4):439–455, October 2004.

[92] Sergey Kitaev and Artem Pyatkin. On representable graphs. *J. Autom. Lang. Comb.*, 13(1):45–54, 2008.

[93] Sergey Kitaev, Pavel Salimov, Christopher Severs, and Henning Úlfarsson. On the representability of line graphs. In Giancarlo Mauri and Alberto Leporati, editors, *Developments in Language Theory*, volume 6795 of *Lecture Notes in Computer Science*, pages 478–479. Springer Berlin Heidelberg, 2011.

[94] Sergey Kitaev, Pavel Salimov, Christopher Severs, and Henning Ulfarsson. Word-representability of line graphs. *Open J. Discrete Math.*, 1(2):96–101, 2011.

[95] Sergey Kitaev and Steve Seif. Word problem of the Perkins semigroup via directed acyclic graphs. *Order*, 25(3):177–194, 2008.

[96] D. J. Kleitman and B. L. Rothschild. Asymptotic enumeration of partial orders on a finite set. *Trans. Amer. Math. Soc.*, 205:205–220, 1975.

[97] Manfred Koebe. On a new class of intersection graphs. In Jaroslav Nešetřil and Miroslav Fiedler, editors, *Fourth Czechoslovakian Symposium on Combinatorics, Graphs and Complexity (Prachatice, 1990)*, volume 51 of *Ann. Discrete Math.*, pages 141–143. North-Holland, Amsterdam, 1992.

[98] Phokion G. Kolaitis, Hans Jürgen Prömel, and Bruce L. Rothschild. K_{l+1}-free graphs: asymptotic structure and a 0–1 law. *Trans. Amer. Math. Soc.*, 303(2):637–671, 1987.

[99] Dénes König. Über Graphen und ihre Anwendung auf Determinantentheorie und Mengenlehre. *Math. Ann.*, 77(4):453–465, 1916.

[100] Nicholas Korpelainen and Vadim Lozin. Two forbidden induced subgraphs and well-quasi-ordering. *Discrete Math.*, 311(16):1813–1822, 2011.

[101] Joseph B. Kruskal. The theory of well-quasi-ordering: A frequently discovered concept. *J. Combinatorial Theory Ser. A*, 13(3):297–305, 1972.

[102] Casimir Kuratowski. Sur le problème des courbes gauches en Topologie. *Fund. Math.*, 15:271–283, 1930.

[103] C. G. Lekkerkerker and J. Ch. Boland. Representation of a finite graph by a set of intervals on the real line. *Fund. Math.*, 51:45–64, 1962/1963.

[104] Vincent Limouzy. Private communication. 2014.

[105] László Lovász. Normal hypergraphs and the perfect graph conjecture. *Discrete Math.*, 2(3):253–267, 1972.

[106] Vadim V. Lozin. On minimal universal graphs for hereditary classes. *Diskret. Mat.*, 9(2):106–115, 1997.

[107] Vadim V. Lozin and Gábor Rudolf. Minimal universal bipartite graphs. *Ars Combin.*, 84:345–356, 2007.

[108] N. V. R. Mahadev and U. N. Peled. *Threshold Graphs and Related Topics*, volume 56 of *Annals of Discrete Mathematics*. North-Holland, Amsterdam, 1995.

[109] Brendan D. McKay and Adolfo Piperno. Practical graph isomorphism II. *Journal of Symbolic Computation*, 60:94–112, 2014.

[110] Alison Miller. Asymptotic bounds for permutations containing many different patterns. *J. Combin. Theory Ser. A*, 116(1):92–108, 2009.

[111] J. W. Moon. On minimal n-universal graphs. *Proc. Glasgow Math. Assoc.*, 7:32–33 (1965), 1965.

[112] Haiko Müller. On edge perfectness and classes of bipartite graphs. *Discrete Math.*, 149(1-3):159–187, 1996.

[113] Peter Nightingale, Özgür Akgün, Ian P. Gent, Christopher Jefferson, and Ian Miguel. Automatically improving constraint models in Savile Row through associative-commutative common subexpression elimination. In Barry O'Sullivan, editor, *Principles and Practice of Constraint Programming - 20th International Conference, CP 2014, Lyon, France, September 8-12, 2014. Proceedings*, volume 8656 of *Lecture Notes in Computer Science*, pages 590–605. Springer, 2014.

[114] P. Perkins. Bases for equational theories of semigroups. *Journal of Algebra*, 11(2):298–314, 1969.

[115] Marko Petkovšek. Letter graphs and well-quasi-order by induced subgraphs. *Discrete Math.*, 244(1-3):375–388, 2002. Algebraic and Topological Methods in Graph Theory (Lake Bled, 1999).

[116] Oliver Pretzel. On graphs that can be oriented as diagrams of ordered sets. *Order*, 2(1):25–40, 1985.

[117] Hans Jürgen Prömel and Angelika Steger. Excluding induced subgraphs: quadrilaterals. *Random Structures Algorithms*, 2(1):55–71, 1991.

[118] Hans Jürgen Prömel and Angelika Steger. Excluding induced subgraphs. III. A general asymptotic. *Random Structures Algorithms*, 3(1):19–31, 1992.

[119] Hans Jürgen Prömel and Angelika Steger. Excluding induced subgraphs. II. Extremal graphs. *Discrete Appl. Math.*, 44(1–3):283–294, 1993.

[120] Heinz Prüfer. Neuer Beweis eines Satzes über Permutationen. *Arch. Math. Phys.*, 27:742–744, 1918.

[121] Neil Robertson and P. D. Seymour. Graph minors. XX. Wagner's conjecture. *J. Combin. Theory Ser. B*, 92(2):325–357, 2004.

[122] Alexander A. Sapozhenko. The Cameron-Erdős conjecture. *Discrete Mathematics*, 308(19):4361 – 4369, 2008.

[123] Edward R. Scheinerman and Jennifer S. Zito. On the size of hereditary classes of graphs. *J. Combin. Theory Ser. B*, 61(1):16–39, 1994.

[124] B.S.W. Schröder. *Ordered Sets: An Introduction.* Springer, Berlin, 2003.

[125] Steve Seif. Monoids with sub-log-exponential free spectra. *J. Pure Appl. Algebra*, 212(5):1162–1174, 2008.

[126] N. J. A. Sloane. The On-line Encyclopedia of Integer Sequences. *Available on-line at http://oeis.org.*

[127] Daniel A. Spielman and Miklós Bóna. An infinite antichain of permutations. *Electron. J. Combin.*, 7:Note 2, 4 pp. (electronic), 2000.

[128] Jeremy P. Spinrad. Nonredundant 1's in Γ-free matrices. *SIAM J. Discrete Math.*, 8(2):251–257, 1995.

[129] Jeremy P. Spinrad. *Efficient graph representations*, volume 19 of *Fields Institute Monographs.* American Mathematical Society, Providence, 2003.

[130] Carsten Thomassen. A short list color proof of Grötzsch's theorem. *Journal of Combinatorial Theory, Series B*, 88(1):189 – 192, 2003.

[131] A. Thue. Über unendliche Zeichenreihen. *Norske Vid. Selsk. Skr. I Math-Nat. Kl.*, 7:1–22, 1906.

[132] Alexander Valyuzhenich. Some properties of well-based sequences. *Journal of Applied and Industrial Mathematics*, 5(4):612–614, 2011.

[133] A.C.M. van Rooij and Herbert S. Wilf. The interchange graph of a finite graph. *Acta Mathematica Academiae Scientiarum Hungaricae*, 16(3-4):263–269, 1965.

[134] Bjorn Winterfjord. Binary strings and substring avoidance. *Master's thesis, CTH and Göteborg University*, 2002.

[135] Mihalis Yannakakis. The complexity of the partial order dimension problem. *SIAM Journal on Algebraic and Discrete Methods*, 3(3):351–358, 1982.

Index

Printed in the United States
By Bookmasters